Star Identification

Guangjun Zhang

Star Identification

Methods, Techniques and Algorithms

国防工业出版社
National Defense Industry Press

Springer

Guangjun Zhang
Beihang University
Beijing
China

ISBN 978-3-662-53781-7 ISBN 978-3-662-53783-1 (eBook)
DOI 10.1007/978-3-662-53783-1

Jointly published with National Defense Industry Press, Beijing, China

Library of Congress Control Number: 2016958496

Printed on acid-free paper

This Springer imprint is published by Springer Nature
The registered company is Springer-Verlag GmbH Germany
The registered company address is: Heidelberger Platz 3, 14197 Berlin, Germany

Preface

Attitude measurement is key and vital for spacecraft. It guarantees accurate orbit entrance and orbit transfer, high-quality performance of spacecraft, reliable space-to-ground communication, high-resolution earth observation, and successful completion of many other missions to be conducted in space. Star sensor is the core component in the autonomous high-quality attitude measurement of in-orbit spacecraft based on the observation of stars. By taking advantage of a star's astronomical information, the star sensor method has the characteristics of good autonomy, high precision, and high reliability and can be widely applicable in space flight (celestial navigation).

Generally speaking, star sensor works in two modes, namely, Initial Attitude Establishment and Tracking. The star sensor enters into the Initial Attitude Establishment Mode when it starts working or when attitude gets lost in space due to unforeseen problems. In this mode, full-sky star identification is needed because there is no available attitude information. Once initial attitude is established, the star sensor enters into Tracking Mode. Full-sky autonomous star identification is key in star sensor technological development, which has encountered many difficulties and therefore it is a focus for research.

Star identification is interdisciplinary and related to astronomy, image processing, pattern recognition, signal and data processing, computer science, and many other fields of study. This book summarizes the research conducted by the author's team for more than ten years in this specific field. There are seven chapters, covering basics in star identification, star cataloging and star image preprocessing, principles and processes of algorithms, and hardware implementation and performance testing. Chapter 1 is a general introduction, covering basics in celestial navigation, with a discussion on star sensor method and star identification, and reviews algorithms used in star identification and the development trends in this field. Chapter 2 deals with the preliminary work in star identification, covering star cataloging, selection of guide stars, processing of a double star, star image simulation, star spot centroiding, and calibration of centroiding error. Chapter 3 is a brief introduction to star identification using triangle algorithms, with a special emphasis on two modified examples, namely angular distance matching and the *P* vector. Chapter 4 focuses on star identification using star patterns, including star identification utilizing radial and cyclic star patterns, by using the log-polar transformation method, also without calibration parameters. Chapter 5 discusses basic principles of star identification using neural networks. Two methods are presented—star identification based on neural networks carried out by using features of star vector matrix and by also using mixed features. Chapter 6 introduces rapid star tracking using star matching between adjacent frames, covering star tracking modes of the star sensor method, different algorithms in star tracking, with simulation results presented and analyzed. Chapter 7, taking RISC CPU as an example, deals with hardware implementation, as well as hardware-in-the-loop simulation testing and field experimentation of star identification.

For many years, the author's research team has obtained the support from Major Research Grants for Civil Space Programs, the National Natural Science Foundation of China, Chinese National Programs for High Technological Research and Development (863 Program), and Aerospace Engineering projects. The author wishes to thank the Department of Science, Technology and Quality Control of the former State Commission of Science, Technology and Industry for National Defense, the National Natural Science Foundation of China, the Department of High and New Technology Development and Industrialization of the Ministry of Science and Technology, and the Shanghai Academy of Spaceflight Technology for their support.

This book is based on many years of research on star identification by the author and his team. The author wants to express his gratitude to the following people in his team—Xinguo Wei, Jie Jiang, Qiaoyun Fan, Xuetao Hao, Jian Yang, Juan Shen, Xiao Li, and many others, who have contributed to much of the work introduced in this book. The author is also indebted to the National Defense Industry Press for including this monograph in its book series on spacecraft and guided missiles.

Citations in the book are given due credit. References are listed so that interested readers know where to look for further information.

Star identification involves a wide range of topics and is related to many research fields. The author does not venture to cover all in this single book and knows clearly the limitations that may exist. Any mistakes, therefore, remain the sole responsibility of the author.

Beijing, China
December 2010

Guangjun Zhang

Contents

1 Introduction . . . 1
1.1 Fundamental Knowledge of Astronomy . . . 2
1.1.1 Characteristics of Stars . . . 2
1.1.2 The Celestial Sphere and Its Reference Frame . . . 3
1.1.3 Star Catalog . . . 5
1.2 Introduction to Celestial Navigation . . . 6
1.2.1 Basic Principles of Celestial Navigation . . . 6
1.2.2 Characteristics of Celestial Navigation and Its Framework . . . 10
1.3 Introduction to Star Sensor Technique . . . 11
1.3.1 Principles of Star Sensor Technique and Its Structure . . . 12
1.3.2 The Current Status of Star Sensor Technique . . . 13
1.3.3 Development Trends in Star Sensor Technique . . . 17
1.4 Introduction to Star Identification . . . 19
1.4.1 Principles of Star Identification . . . 19
1.4.2 The General Process of Star Identification . . . 20
1.4.3 Evaluation of Star Identification . . . 23
1.5 Star Identification Algorithms and Development Trends . . . 24
1.5.1 Subgraph Isomorphism Algorithms . . . 25
1.5.2 Star Pattern Recognition Class Algorithms . . . 27
1.5.3 Other Algorithms . . . 30
1.5.4 Development Trends of Star Identification Algorithms . . . 32
1.6 Introduction to the Book Chapters . . . 33
References . . . 33

2 Processing of Star Catalog and Star Image . . . 37
2.1 Star Catalog Partition . . . 37
2.1.1 Guide Star Catalog . . . 38
2.1.2 Current Methods in Star Catalog Partition . . . 39
2.1.3 Star Catalog Partition with Inscribed Cube Methodology . . . 41

2.2 Guide Star Selection and Double Star Processing 43
2.2.1 Guide Star Selection 43
2.2.2 Double Star Processing 45
2.3 Star Image Simulation 48
2.3.1 The Image Model of the Star Sensor 49
2.3.2 The Composition of the Digital Star Image 53
2.4 Star Spot Centroiding 55
2.4.1 Preprocessing of the Star Image 56
2.4.2 Centroiding Methods 58
2.4.3 Simulations and Results Analysis [17] 61
2.5 Calibration of Centroiding Error 65
2.5.1 Pixel Frequency Error of Star Spot Centroiding 66
2.5.2 Modeling of Pixel Frequency Error 69
2.5.3 Calibration of Pixel Frequency Error 69
References 71

3 Star Identification Utilizing Modified Triangle Algorithms 73
3.1 Current Triangle Algorithms 74
3.1.1 Basic Principles of the Triangle Algorithm 74
3.1.2 Problems with the Triangle Algorithm 76
3.2 Modified Triangle Algorithm Utilizing the Angular Distance Matching Method 78
3.2.1 Star Pairs Generation and Storage 78
3.2.2 Selection of Measured Triangles 80
3.2.3 Identification of Measured Triangles 82
3.2.4 Process of Verification 83
3.2.5 Simulations and Results Analysis 87
3.3 Modified Triangle Algorithm Utilizing the *P* Vector 92
3.3.1 *P* Vector Generation 93
3.3.2 Construction of a Guide Database 97
3.3.3 Matching and Identification 99
3.3.4 Simulations and Results Analysis 101
References 105

4 Star Identification Utilizing Star Patterns 107
4.1 Introduction to the Grid Algorithm 108
4.1.1 Principles of the Grid Algorithm 108
4.1.2 Deficiencies of the Grid Algorithm 109
4.2 Star Identification Utilizing Radial and Cyclic Star Patterns 111
4.2.1 Star Patterns Generation and Storage 111
4.2.2 Process of Identification 114
4.2.3 Simulations and Results Analysis 117
4.3 Star Identification Utilizing the Log-Polar Transformation Method 123

4.3.1 Principles of Log-Polar Transformation 123
4.3.2 Star Pattern Generation Utilizing the Log-Polar Transformation Method 124
4.3.3 Star Pattern String Coding and Recognition 127
4.3.4 Simulations and Results Analysis 133
4.4 Star Identification Without Calibration Parameters 136
4.4.1 Influence of Intrinsic Parameters of a Star Sensor on Star Identification 137
4.4.2 Extraction of Feature Patterns Independent of Calibration Parameters 139
4.4.3 Matching and Identification 141
4.4.4 Simulations and Results Analysis 144
References 150

5 Star Identification Utilizing Neural Networks 153
5.1 Introduction to Neural Networks 154
5.1.1 Basic Concepts of Neural Networks 154
5.1.2 Basic Characteristics of Neural Networks 155
5.1.3 Basic Principles of Neural Networks 157
5.2 Star Identification Utilizing Neural Networks Based on Features of a Star Vector Matrix 159
5.2.1 Self-organizing Competitive Neural Networks 159
5.2.2 Extraction and Storage of Guide Star Patterns 161
5.2.3 Construction of Self-organizing Competitive Neural Networks 164
5.2.4 Simulations and Results Analysis 167
5.3 Star Identification Utilizing Neural Networks Based on Mixed Features 171
5.3.1 Construction of Competitive Neural Networks 171
5.3.2 Extraction and Identification of Star Pattern Vectors 173
5.3.3 Simulations and Results Analysis 175
References 176

6 Rapid Star Tracking by Using Star Spot Matching Between Adjacent Frames 177
6.1 Tracking Mode of the Star Sensor 177
6.1.1 Principles of Star Tracking 177
6.1.2 Characteristics of Star Tracking 179
6.1.3 Current Star Tracking Algorithms 180
6.2 Rapid Star Tracking Algorithm by Using Star Spot Matching Between Adjacent Frames 181
6.2.1 Basic Principles of Star Tracking Algorithm 181
6.2.2 Guide Star Indexing by Using Partition of Star Catalog . . . 185
6.2.3 Threshold Mapping 185

6.2.4 Sorting Before Matching and Identification. 187
6.2.5 Star Spot Position Prediction. 188
6.3 Simulations and Results Analysis 190
6.3.1 Selecting Star Tracking Parameters 190
6.3.2 Influence of Star Position Noise on Star Tracking. 193
6.3.3 Influence of Star Sensor's Attitude Motion on Star Tracking 195
6.3.4 Speed of Star Tracking 197
References. 198

7 Hardware Implementation and Performance Test of Star Identification 199
7.1 Implementation of Star Identification on RISC CPU 200
7.1.1 Overall Structural Design of RISC Data Processing Circuit. 200
7.1.2 Selection of Primary Electronic Components 201
7.1.3 Hardware Design of RISC Data Processing Circuit. 203
7.1.4 Software Design of RISC Data Processing Circuit 206
7.2 Hardware-in-the-Loop Simulation Test of Star Identification 208
7.2.1 Test System Configuration and Test Methods. 209
7.2.2 Function Test of Star Identification and Star Tracking 210
7.2.3 Time Taken in Full-Sky Star Identification 212
7.2.4 Update Rate of Attitude Data 213
7.3 Field Experiment of Star Identification. 215
7.3.1 Manual Star Identification by Using Skymap 216
7.3.2 Function Test of Star Identification and Star Tracking 220
References. 223

Introduction

Star identification is the essential guarantee for the working performance of star sensors and one key step for celestial navigation. This book summarizes the research findings by the author's team in the area of star identification for more than ten years, with a systematic introduction of the principles of star identification, as well as the general methods, key techniques, and practicable algorithms. Topics covered include fundamental knowledge of star sensor and celestial navigation, processing of the star catalog and star image, star identification methods undertaken by using modified triangle algorithms, star identification utilizing star patterns, star identification by using neural networks, rapid star tracking by using star spot matching between adjacent frames, and hardware implementation and performance testing of star identification.

This book can be used as a course book for senior undergraduate students and postgraduate students majoring in information processing, computer science, artificial intelligence, aeronautics and astronautics, automation and instrumentation. Moreover, this book can also be used as a reference for people engaged in pattern recognition and other related research areas.

Chapter 1
Introduction

Navigation systems are vital and indispensable for spacecraft. The main task of a navigation system is to guide a spacecraft to its destination following predetermined routes with the required precision and within the given time. For this purpose, the system should provide accurate navigation parameters, including azimuth (i.e., horizontal attitude and course), velocity, position, etc. Since these parameters can be obtained using various physical principles and techniques, there exist different types of navigation systems [1, 2], e.g., radio navigation systems, inertial navigation systems, GPS navigation systems, terrain matching navigation systems, scene matching navigation systems, celestial navigation systems, and integrated navigation systems, which are an integration of multiple navigation systems.

Based on known coordinate positions and motion rules of celestial bodies, celestial navigation uses the astronomical coordinates of an observed object to determine the geographical position and other navigation parameters of a spacecraft. Celestial navigation is not applicable to aircraft within the Earth's atmosphere as they are subject to climate conditions. However, for crafts entering thin air or navigating at over 8000 m above the ground, it is highly reliable to utilize information provided by a celestial navigation system. Different from other navigation technologies, celestial navigation is autonomous and requires no ground equipment. Free from interference from artificially or naturally formed electromagnetic fields, it radiates no energy externally. In addition, the system is well concealed and highly precise in the determination of the attitude, orientation, and position of spacecraft, and the time of navigation is not linked to positioning errors. In general, celestial navigation is very promising in terms of its applications.

In this chapter, fundamental astronomic knowledge and principles of celestial navigation are introduced first. The second part summarizes technologies for star sensors and star identification. Star identification algorithms as well as their development trends are also explained in this chapter. Then each chapter of the book is briefly introduced.

G. Zhang, *Star Identification*, DOI 10.1007/978-3-662-53783-1_1

1.1 Fundamental Knowledge of Astronomy

Celestial navigation works on the basis of celestial information obtained by celestial sensors, the prerequisite of which is prior knowledge about the characteristics and motion rules of celestial bodies. Thus, fundamental knowledge of the characteristics and motion rules of celestial bodies is crucial for studies of celestial navigation. In this section, astronomic knowledge relevant to star sensors and star identification is introduced.

1.1.1 Characteristics of Stars

Celestial navigation observes celestial bodies. Among all the objects observed, stars are the most important type. Hence, it is necessary to acquire a basic understanding of the characteristics of stars, which are summarized as follows [3, 4]:

(1) Distance of stars. Stars are quite remote from the Earth. Except for the Sun, the nearest star to Earth is Centaurus, which is 4.22 light years away. Therefore, in celestial navigation, stars can be regarded as celestial bodies at infinite distances.

(2) Velocity of stars. Stars, also known as fixed stars, are usually considered to be stationary. Actually, stars are constantly moving at high speeds in space. The velocity of a star can be decomposed into radial velocity and tangential velocity. The former refers to the component measured along the observer's line of sight (positive when the observed object is moving away from the observer and negative when it is moving towards the observer), while the latter is the component measured along the line perpendicular to the observer's line of sight. Tangential velocity usually shows up as displacement of stars in the celestial sphere. Our concern is usually this displacement of stars, also known as proper motion. The velocity of stars' proper motion is generally less than $0.1''$ per year. So far, only around 400 stars have been observed to be moving more than $1''$ a year.

(3) Brightness of stars. As an inherent characteristic, stars emit visible light on their own. The brightness of a star refers to its apparent brightness observed from the Earth, which is subject to both its luminosity (related to its temperature and size) and the distance between the star and the Earth. In astronomy, the degree of brightness of a star is evaluated with a unit of measurement called star magnitude (also known as visual magnitude, Mv). The lower the magnitude is, the brighter the star is. A decrease of one in magnitude represents an increase in brightness of 2.512 times. A star of 1 Mv is approximately 100 times brighter than one of 6 Mv. Two stars, Aldebaran (Alpha (α) Tauri) and Altair (Alpha (α) Aquilae), were originally assigned as the standard stars for 1.0 Mv in astronomy. Later, Vega (Alpha (α) Lyrae) was

Table 1.1 Visual magnitude of common celestial bodies

Celestial body	Mv
Sun	−26.5
Moon (full moon)	−12.5
Venus (brightest moment)	−4.0
Sirius	−1.46
Polaris	2.02

adopted as the standard star for 0.0 Mv and all other stars' Mv were referenced to this. Table 1.1 illustrates the visual magnitude of some common celestial bodies. Stars of 6 Mv or brighter can be seen with the naked eye. Through astronomical telescope, stars of 10 Mv or brighter are observable. The Hubble space telescope enables the observation of stars of up to 30 Mv.

(4) Size of stars. Stars vary significantly in size. However, when observed from the Earth, their field angles are far smaller than $1''$, making it reasonable to treat a star as an ideal point light source.

To sum up, in celestial navigation, stars can be generally considered as nearly stationary point light sources with certain spectral characteristics at infinite distances.

1.1.2 The Celestial Sphere and Its Reference Frame

To better describe of the azimuth of stars, it is necessary to construct a reference frame in order to describe the position of a star at a certain moment with a set of coordinate values. This frame is called the celestial coordinate system. Below are some astronomical definitions and concepts relevant to the system [5]:

(1) Celestial sphere. In astronomy, the sky is pictured as a huge sphere, named the celestial sphere, matching people's intuitive perception of the sky, as shown in Fig. 1.1. The celestial sphere is an imaginary sphere with infinitely large radius, concentric with Earth. Its infinitely large radius makes any finite distance negligible. Hence, any random spot on the surface or in a certain region of the Earth can be treated as the center of the celestial sphere.

(2) Celestial axis and celestial poles. The celestial axis is the imaginary straight line that goes through the center of the celestial sphere and is in parallel with the Earth's axis of rotation, which is shown as PP' in Fig. 1.1. The intersections of the axis and the sphere are celestial poles. P is the North celestial pole and P' is the South celestial pole.

(3) Celestial equator and its plane. The celestial equator, illustrated as $QEQ'W$ in Fig. 1.1, is the line that goes through the center of the celestial sphere and is perpendicular to the axis. The celestial equator plane is the plane where the equator lies.

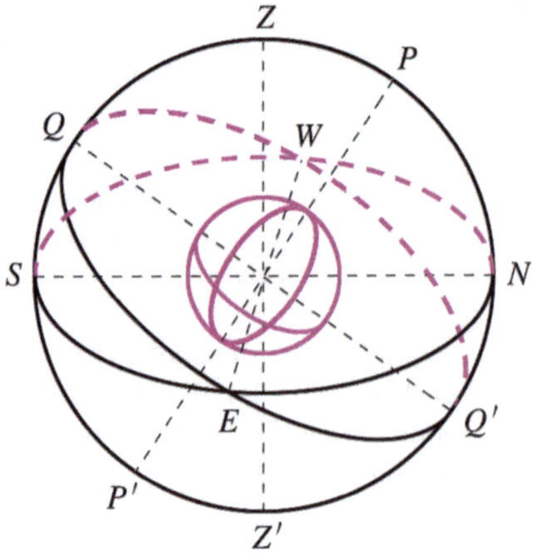

Fig. 1.1 Celestial sphere

(4) Hour circle. Hour circles are the big circles that meet the celestial sphere by passing through the celestial poles.

(5) Ecliptic and ecliptic pole. The mean plane of the Earth's orbit around the Sun is the ecliptic plane. Its intersection with the celestial sphere is a large circle, i.e., the ecliptic. The ecliptic poles refer to the points on the celestial sphere where the sphere meets the imaginary line that passes through the celestial center and is perpendicular to the ecliptic plane. The obliquity of the ecliptic, i.e., the angle between the ecliptic plane and the equator plane, is 23° 27′.

(6) Vernal equinox. The equator and the ecliptic intersect at two opposite points. Vernal equinox γ is the point at which the ecliptic crosses the equator moving northward.

(7) Celestial coordinate system. The second equatorial coordinate system is defined as a coordinate system with the celestial equator as its fundamental circle (or abscissa circle), the hour circle passing through vernal equinox γ as its primary circle and the vernal equinox as its principal point. In astronomy, the second equatorial coordinate system is also called the right ascension coordinate system, and also known as the celestial coordinate system. The position of a celestial body is determined by its right ascension and declination in the system. As Fig. 1.2 shows, $Q\gamma Q'$ refers to the plane of the celestial equator, while α and δ stand for the right ascension and declination of the celestial body, respectively. It is stipulated that right ascension is measured counterclockwise (opposite to the direction of diurnal motion), ranging from 0° to 360°. Declination, on the other hand, is measured from the celestial equator towards the north and the south, ranging from 0 to +90° and 0 to −90°, respectively. The position of a star is generally described by giving these two coordinates in the celestial coordinate system.

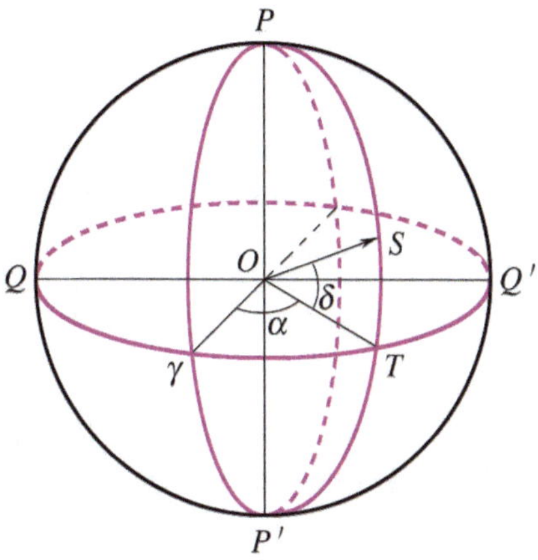

Fig. 1.2 Celestial coordinate system

1.1.3 Star Catalog

A star catalog is an astronomical catalog that lists stars and their data according to different needs [3]. Usually, a star catalog records the position (marked by right ascension and declination), proper motion, brightness (measured by star magnitude), color, distance, and many other details of a star. It serves as the foundation and criterion for star identification and attitude determination. Frequently used star catalogs include the U.S. Smithsonian Astrophysical Observatory Catalogue (SAO), Hipparcos Catalogue (HIP or HP), Henry Draper Catalogue (HD), Bright Star Catalogue (BS or BSC), etc. The SAO J2000 (epoch = J2000) compiled by the U.S. Smithsonian Astrophysical Observatory, recording around 250,000 stars brighter than 17 Mv, is adopted as the standard catalog internationally [6].

The most useful data for astronomical navigation are the position and brightness of stars. The position of a star, i.e., the projection of a star onto the celestial sphere, is further decomposed into mean position, true position, and apparent position. The position of a star recorded in the standard star catalog refers to its mean position at the standard epoch (J2000). Using the mean position at the standard epoch and the precession and proper motion from the standard epoch to the current mid-year, one can calculate the mean position of a star in this particular mid-year. The mean position on a particular day can then be gained by adding the mean position in mid-year to the precession and proper motion of the specified day. Furthermore, the nutation and mean position on a particular day can be summed up, giving the true position of a star. The apparent position of a star refers to the star's coordinates in the celestial coordinate system when observed. It can be obtained when the solar coordinate system is converted into the Earth coordinate system, i.e., when the aberration of light is taken into account. For simplicity, astronomers employ the coordinate sets of right ascension and declination of the standard catalog (namely the mean position at the standard epoch). In this way, the reference coordinate system can be treated as the mean coordinate system of the standard epoch (a coordinate system with mean equinox and mean equator as its coordinate axes). Therefore, the attitude calculated is in some sense based on the mean coordinate system of the standard epoch.

For the convenience of subsequent star identification and attitude calculation, the right ascension and declination of stars are usually regarded and recorded as:

$$\begin{pmatrix} x \\ y \\ z \end{pmatrix} = \begin{pmatrix} \cos\alpha\cos\delta \\ \sin\alpha\cos\delta \\ \sin\delta \end{pmatrix} \tag{1.1}$$

This recording approach avoids trigonometric operations in star identification and attitude calculation. Therefore, it saves time.

1.2 Introduction to Celestial Navigation

In the past, by observing stars, people utilized their relatively stationary position and the predictable motion of Earth to navigate. Taking the horizon as a local horizontal reference, an observer in the Northern Hemisphere can estimate the local latitude by looking up at the Big Dipper. In fact, this is the navigating approach used by ancient mariners. However, except for a local horizontal reference, other factors including the exact time of observation (year, month, date, and moment) and the ephemeris that demonstrates the position of stars are also required for the estimation of both latitude and longitude of a location. Navigators of cross-ocean flights originally used sextants with bubble levels to manually measure the line-of-sight angle (also known as visual angle), which is the angle between the local perpendicular and the line from a star to the observer's eye. Taking advantage of the line-of-sight angles of two or more stars, the ephemeris of stars and the exact time of observation, navigators could calculate the local latitude and longitude.

Thanks to progress in optoelectronic and computer technology, and especially the emergence of CCD (Charge-Coupled Device) and CMOS (Complementary Metal–Oxide–Semiconductor) imaging devices, celestial navigation technology has entered a new stage of development. This technology has been widely employed in satellites, space shuttles, intermediate-range ballistic missiles, and other spacecraft. In this section, the fundamental principles and composition of celestial navigation systems are briefly introduced.

1.2.1 Basic Principles of Celestial Navigation

The main task of celestial navigation is to determine the attitude and position of a spacecraft. This section introduces the basic principles of celestial navigation [7].

(1) Attitude Determination Using Two Vectors

Parallel lights emitted by stars are imaged on the focal plane through the optical system. The central position of a star image (P_x, P_y) can be determined using a centroid approach. According to imaging geometry and on the basis of the central position of a star image, the direction of a starlight vector relative to the coordinates of a spacecraft can be obtained. By referring to a star catalog, the position of a star in the celestial coordinate system can be described by its right ascension and declination. Hence, based on two or more stars, the attitude matrix of a spacecraft with respect to the celestial coordinate system can be calculated. The detailed algorithm of attitude determination using two vectors is shown in Fig. 1.3.

Denote the image space coordinate system as S_m, the right ascension coordinate system as S_r and the transformation matrix as $\mathbf{C}_{mr}$. Suppose there are two starlight vectors, and their direction vectors in S_m and S_r are defined as W_1, W_2, and U_1, U_2, respectively.

A reference coordinate system, S_c, is constructed on the basis of these two measurement vectors. Its orthonormal coordinate basis in S_m coordinate system is

$$a = W_1, \quad b = (W_1 \times W_2)/|W_1 \times W_2|, \quad c = a \times b \tag{1.2}$$

The attitude transformation matrix from S_c to S_m is

$$\mathbf{C}_{cm} = \begin{pmatrix} a^{\mathrm{T}} \\ b^{\mathrm{T}} \\ c^{\mathrm{T}} \end{pmatrix} \tag{1.3}$$

Similarly, the orthonormal coordinate basis of S_c in S_r coordinate system is

$$A = U_1, \quad B = (U_1 \times U_2)/|U_1 \times U_2|, \quad C = A \times B \tag{1.4}$$

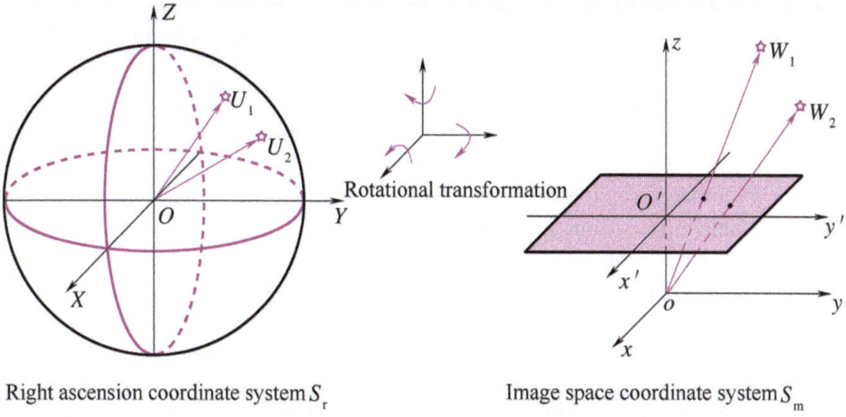

Fig. 1.3 Illustration of attitude determination using two vectors

The attitude transformation matrix from S_c to S_r is

$$\mathbf{C}_{cr} = \begin{pmatrix} A^{\mathrm{T}} \\ B^{\mathrm{T}} \\ C^{\mathrm{T}} \end{pmatrix} \tag{1.5}$$

Since $S_c = \mathbf{C}_{cm}S_m = \mathbf{C}_{cr}S_r$, $S_m = \mathbf{C}_{mr}S_r$, it can be deduced that $\mathbf{C}_{mr} = \mathbf{C}_{cm}^{-1}\mathbf{C}_{cr}$. In accordance with this equation, the attitude transformation matrix of the image space coordinate system relative to the right ascension coordinate system can be calculated. Then the attitude information of a spacecraft can be gained.

(2) Principle of Positioning

To determine the spatial position of an object, a type of optical observation data, i.e., the direction of lines of sight of several nearby celestial bodies, is needed. The positions of these nearby celestial bodies should have been clarified and measured in a known inertial reference frame. This inertial reference frame can be constructed using either two random noncollinear star lines (lines of sight), or three random noncoplanar star lines or platform coordinates. Clearly, only by measuring the position of nearby celestial bodies, spatial positioning can be endowed with positional significance. As regards the angle that should be measured in order to determine the position of a spacecraft, this is essentially the included angle between lines of sight of a nearby celestial body, because the angle between a star (inertial frame) and the center of a planet (nearby celestial body), for instance, changes as the spacecraft moves.

On the other hand, the angle between the star lines at two different positions does not change with different measurements. Hence, changes in angle can represent changes in position. Of course, the position of a star may vary slightly during a spacecraft's interstellar voyage, calling for correction. For near-Earth navigation, however, this deviation may be ignored since it is too tiny to affect the high accuracy of positioning.

In celestial positioning, as shown in Fig. 1.4, an imaging system is utilized to measure the angle between a star and a planet (nearby celestial body). The spacecraft's position can then be obtained on the basis of a spatial cone. Since the degree of the angle between the lines of sight of a star and a nearby celestial body is constant, this cone can be worked out, with the spacecraft being located on the surface of this cone. If one measures a second star and the same nearby celestial body in a similar way, another cone can be obtained. These two cones share one apex and their intersections are two lines, as shown in Fig. 1.5. It can be inferred that the spacecraft is on one of these lines. By distinguishing or observing a third star, one can clarify the specific line on which the spacecraft lies.

To find out the precise position of the spacecraft on the line, a second nearby celestial body should be selected. Prior knowledge of the position vector of the second body toward the first one is also needed. A third cone, gained by measuring the second nearby celestial body and the third star, intersects with the former two cones at two points, *a* and *c*, as shown in Fig. 1.6. By again following the above

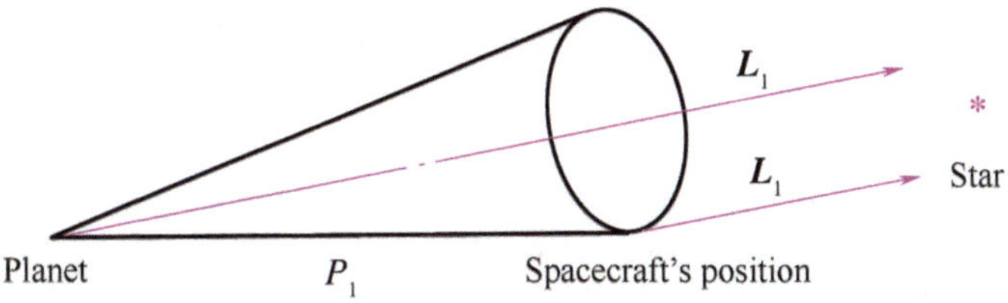

Fig. 1.4 The cone of positioning determined through single star observation

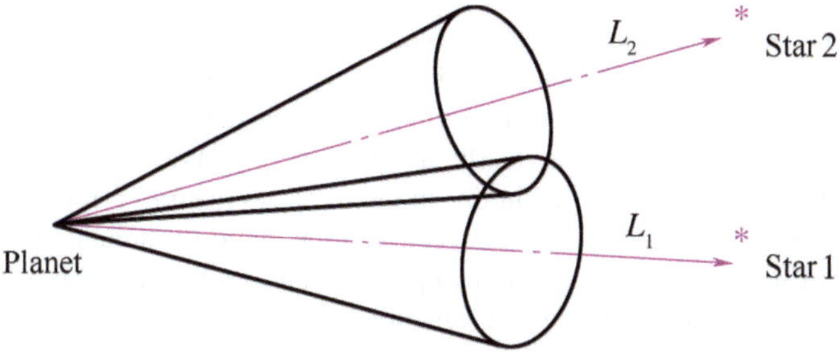

Fig. 1.5 Two position lines determined through double star observation

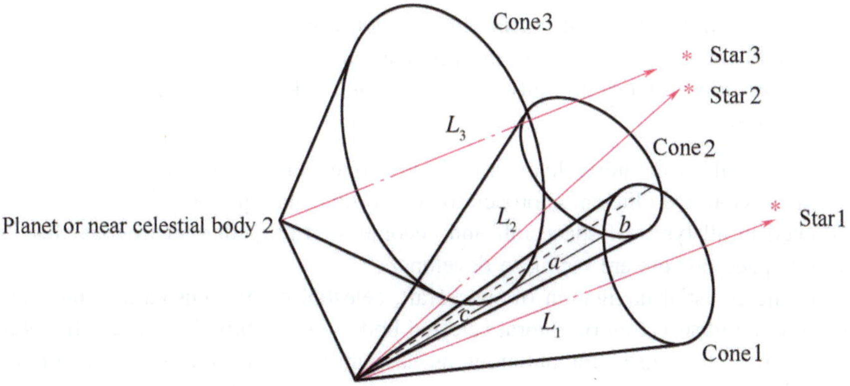

Fig. 1.6 Illustration of complete positioning

approach, selecting a real point from the two intersections of the three cones, the position of the spacecraft relative to any nearby celestial body can be expressed.

As stated above, a star catalog and ephemeris information of at least two nearby celestial bodies (planets) are required for celestial positioning. Such information is needed by all kinds of positioning technologies (including technology making use of two nearby celestial bodies, and those utilizing line-of-sight technology or landmark tracking). Readers interested in the topic may refer to relevant literature for detailed algorithms related to positioning.

1.2.2 Characteristics of Celestial Navigation and Its Framework

Celestial navigation determines the position and attitude of a craft by observing natural celestial bodies (the Sun, stars, planets, the Moon, etc.). Its merits are summarized as follows [8]:

① Passive measurement and autonomous navigation. Celestial navigation, adopting celestial bodies as navigation beacons, does not depend on other external information. It passively receives radiation or reflected light from celestial bodies. Hence, celestial navigation is completely autonomous.
② High precision in measurement and no accumulated error. Celestial navigation works on the basis of observed information of celestial bodies. Once the motion rules of celestial bodies are clearly and accurately attained, celestial navigation can be highly precise. In addition, in celestial navigation, errors do not accumulate with time. Because of this, celestial navigation is suitable for situations that call for extended autonomous operation and high navigating precision.
③ High interference resistance and reliability. Celestial radiation includes a whole range of electromagnetic bands, covering X-ray, ultraviolet, visible light, and infrared. Thus, it enjoys a high resistance to interference. Moreover, celestial navigation adopts celestial bodies as its navigation beacons. The spatial motion of celestial bodies, which is not subject to human disturbances, guarantees the perfection and reliability of the information used for celestial navigation.
④ Simultaneous provision of information on position and attitude. Celestial navigation provides not only the position and velocity of a spacecraft, but also its attitude.

As a result of the unrivaled merits listed above, celestial navigation has become the most vital and efficient approach to autonomous navigation. It has been widely adopted in all types of spacecraft and become a cutting-edge technology that the world space powers are racing to develop.

In the celestial navigation of spacecraft, celestial object sensors are frequently used in the observation of natural celestial bodies. The azimuth of a celestial body observed is then used for autonomous navigation. In accordance with different sensitive objects, celestial object sensors are divided into star sensors, sun sensors, earth sensors, lunar sensors, and others. Among all these, star sensors are most widely used. They adopt star vectors as their reference vectors and guarantee attitude measurement with high precision.

Celestial navigation systems based on star sensors usually consist of a star sensor, a computer, an information processor, a standard time generator, and an inertial platform as a tracking platform. Current technology integrates the computer, information processor, and standard time generator into the star sensor, making them integral parts of the sensor. It can be concluded from the above that the star sensor is the major component of a celestial navigation system.

1.3 Introduction to Star Sensor Technique

Precise attitude information lays the foundation for the autonomous navigation of spacecraft, and serves as the most critical component in a spacecraft's attitude control system. To determine the attitude of a spacecraft, a reliable reference frame (e.g., an inertial space, the Sun, the Earth, or a star) is usually selected first. According to the observed changes with respect to the reference frame, changes in the attitude of a spacecraft can then be deduced. The attitude-measuring component is usually called an attitude sensor. A gyroscope is a kind of attitude sensor with an inertial space as its reference frame. It displays outstanding dynamic performance and relatively high accuracy in instantaneous attitude measurement. However, due to its large drift over long voyages, other attitude sensors are needed for correction. Other widely used attitude sensors include Earth sensors (horizon sensors), Sun sensors, GPS, and magnetometers. The precision of attitude measurement of these sensors is relatively low because of their less accurate reference and measurement vectors. Since the reference vector of these sensors is related to the orbital position of a spacecraft, a kinetic equation should be used to estimate the orbit of the spacecraft. Moreover, these sensors are usually only used to estimate the attitude in one direction. Hence, multiple sensors have to be utilized in order to obtain a three-axis attitude.

Star sensor technology offers a brand new approach to measuring the attitude of spacecraft. It adopts the star coordinate system as its reference frame. Since the spatial position of stars can be considered stationary in the reference frame, and the measurement of starlight vectors is highly precise, star sensors can measure the attitude of a craft quite precisely (up to arc-second level).

The accuracy of attitude measurement of several common attitude sensors is shown in Table 1.2 [9]. Because the attitude measurement of star sensors is irrelevant to the orbits of spacecraft and stars can be spotted everywhere, star sensors are applicable in various situations, including deep space exploration. In addition, being highly reliable and light in weight, star sensors consume relatively little power and work in multiple modes. They operate autonomously, independently outputting a three-axis attitude without relying on other attitude sensors. With all these merits, star sensor technology has become an extraordinarily high-performing spatial attitude-sensitive technology and has been increasingly recognized and widely applied in spacecraft.

Table 1.2 Comparison of the attitude measurement accuracy of common attitude sensors

	Reference frame	Attitude measurement accuracy
Earth sensor	Horizon	$6'$
Sun sensor	Sun	$1'$
Magnetometer	Geomagnetism	$30'$
Star sensor	Star	$1''$

This section first introduces the principles and structure of star sensors. Both the current situation and future development of star sensor technology is then discussed.

1.3.1 Principles of Star Sensor Technique and Its Structure

The operating principle of star sensors is summarized as follows: First, an image sensor (CCD or CMOS) captures an image of the boresight pointing to the night sky. The image is then processed by a signal processing circuit, and information on the position (and brightness) of the stars is extracted. Through a star pattern identification algorithm, the corresponding match of the stars measured is found in the guide star database. Consequently, on the basis of the direction vectors of matched star pairs, the three-axis attitude of the star sensor is calculated, determining the spatial attitude of a spacecraft. A typical operation principle of star sensors is shown in Fig. 1.7.

Defining the direction vector of the measured star in the star sensor coordinate system as $\mathbf{w}$, and the direction vector of the corresponding guide star in the celestial coordinate system as $\mathbf{v}$, then

$$\mathbf{w} = \mathbf{A}\mathbf{v} \tag{1.6}$$

Here, $\mathbf{A}$ stands for the attitude transformation matrix from a celestial coordinate system to a star sensor coordinate system. As an orthogonal matrix, $\mathbf{A}$ satisfies

$$\mathbf{A}^{\mathrm{T}}\mathbf{A} = \mathbf{I} \tag{1.7}$$

When two or more stars in the measured star image have been correctly identified (that is, when corresponding guide stars have been found), the attitude transformation matrix $\mathbf{A}$ can be calculated. The detailed process of computing the attitude matrix according to the measurement vectors of two stars is introduced in detail in Sect. 1.2.1.

Star sensors integrate technologies from optics, mechanics, electronics, image processing, embedded computing, and so on. As shown in Fig. 1.8, a star sensor consists of a baffle, a lens, an image sensor and its circuit board, a signal processing circuit, a housing structure, an optical cube, and some other components.

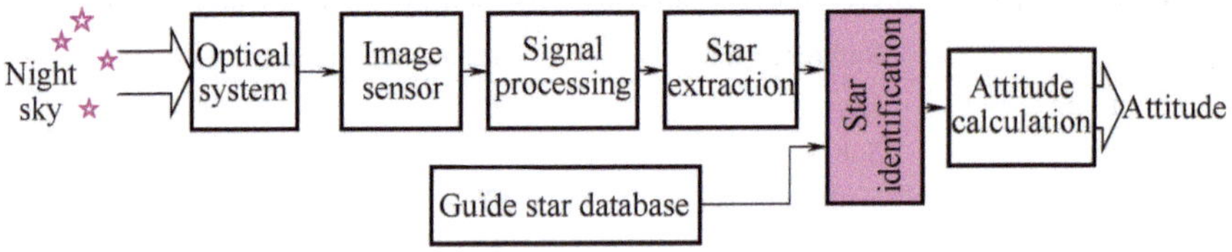

Fig. 1.7 Operation principle of star sensors

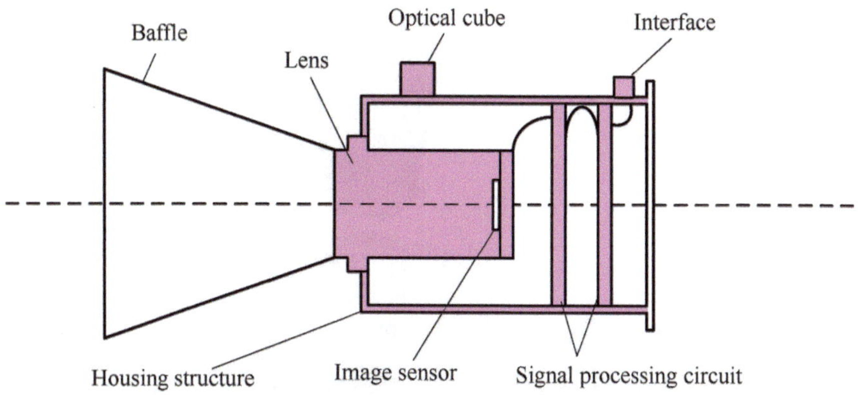

Fig. 1.8 Star sensor structure

The baffle is designed to eliminate the external stray light that shines into the image sensor in order to reduce the background noise of the captured star image. Stray light, including sunlight, earthshine, and so on, can considerably interfere with the star positioning and star identification programs of a star sensor.

The lens of a star sensor images stars at an infinite distance onto the focal plane of the image sensor. The baffle, together with the lens, constitutes the optical system of a star sensor.

As the key component of a star sensor, an image sensor transforms optical signals into electrical ones. Frequently used image sensors can be grouped into two categories, CCD and CMOS. The subsequent signal processing circuit is responsible for the image sensor's imaging drive, timing control, star positioning and star identification, and so on. The circuit finally outputs a three-axis attitude.

There are generally two types of external interfaces for a star sensor: a power interface and a communication interface. The former provides power for the proper functioning of a star sensor, and the latter provides the data communication between a star sensor and the system.

The optical alignment cube fixed to the housing structure facilitates the conversion of the measuring benchmark when the star sensor is being mounted and aligned on a spacecraft.

The internal structure of the ASTRO APS star sensor by Jena-Optronik, a German Company, is shown in Fig. 1.9 [10].

1.3.2 *The Current Status of Star Sensor Technique*

Since the mid-twentieth century, star sensor technology has experienced four stages of development, i.e., early-stage star sensors, first-generation (CCD) star sensors, second-generation CCD star sensors, and CMOS star sensors [11–13]. Currently,

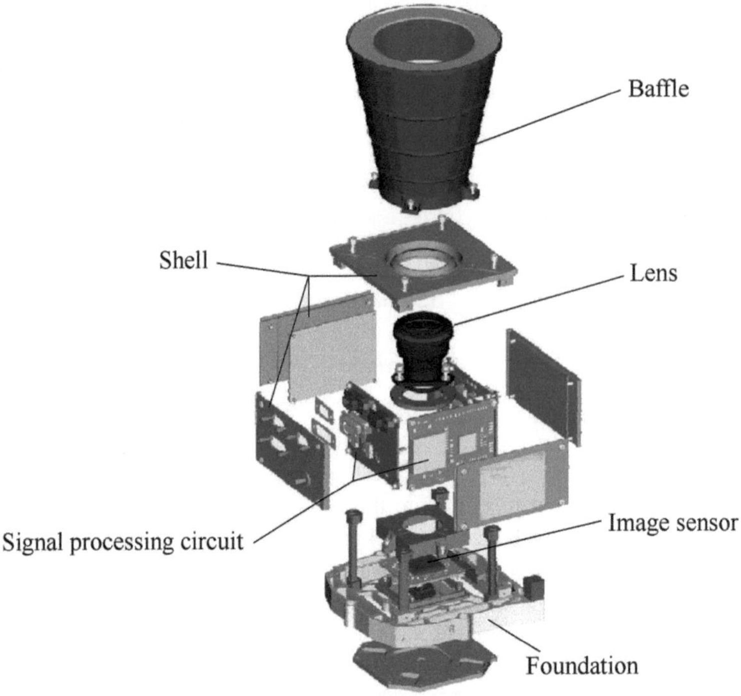

Fig. 1.9 Internal structure of ASTRO APS star sensor

star sensor technology is undergoing a transition from second-generation CCD star sensors to CMOS star sensors.

Early star sensors adopted photomultiplier tubes as their detecting elements. Later, photomultiplier tubes were replaced by dissector tubes. Between the 1960s and 1970s, star sensors with dissector tubes were widely employed in lunar orbiters, the *Apollo* program, Small Astronomy Satellites (SAS-C), International Ultraviolet Explorer (IUE), High Energy Astronomical Observatory (HEAO1,2,3) and many other missions. However, the use of dissector tubes, analog devices that use focus and bias circuits, encountered many formidable problems in practical application. When an accuracy of less than 30 arc seconds is required, dissector tubes can meet this requirement with the assistance of rigorous calibration technology. Nonetheless, over a 2° or larger field of view (FOV), with errors remaining at 1 arc second, the simulated stability rate of dissector tubes must be kept lower than 0.04%. It is almost impossible for a dissector tube to reach this. In addition, the utilization and development of dissector tube technology is also restricted by its size, weight, magnetic effect, high voltage breakdown, and other factors.

In the 1970s, the emergence of CCD (Charge-Coupled Device) and CID (Charge-Injection Device) technologies tremendously accelerated the development of star sensors. The world's first star tracker based on a CCD array image sensor, STELLAR, was developed by the U.S. Jet Propulsion Laboratory (JPL) during the early 1970s. Equipped with 100 × 100-pixel CCD from Fairchild Semiconductor Company and an 8080 microprocessor by INTEL, STELLAR is capable of tracking as many as 10 stars simultaneously with a precision of around 7 arc seconds over a 3° optical FOV. CCDs are small in size, light in weight, low in power consumption, and high in reliability. Thanks to these merits, they have been used extensively in star sensors. With defocusing and centroiding technologies, stars can be precisely located at a sub-pixel level, significantly improving the accuracy of measured star vectors. Successively, U.S. companies such as BALL, TRW, and HDOS developed CCD star trackers with a larger FOV and pixel arrays. These early-stage CCD star sensors (or star trackers) are called first-generation star sensors. This generation is characterized by its accuracy ranging from 100 arc seconds (large FOV) to 3 arc seconds (small FOV), but lacks autonomous star identification and attitude calculation functions.

With the appearance of high-speed microprocessors and large capacity memory, scientists started to develop second-generation star sensors with autonomous star identification and attitude calculation functions. The features of second-generation star sensors are summarized as follows:

① Utilizing a built-in star catalog, second-generation star sensors can autonomously identify stars and solve lost-in-space problems without external processors or external input of initial attitude information;
② The FOV becomes larger and more stars can be observed in it. These features make it possible to realize full-sky autonomous star identification;
③ Second-generation star sensors can directly export attitude information with respect to the inertial reference frame.

With a larger FOV, second-generation star sensors can meet the demand on the number of stars for star identification by detecting only the brighter ones. In addition, the sensor can independently establish initial attitude without relying on external devices. Therefore, lost-in-space problems can be solved and the sensor can navigate autonomously.

Since the 1970s, researchers have been active in studying and developing star sensors. Star sensor technology has been increasingly and widely applied in earth observation, lunar observation, planetary observation, interstellar communication, spacecraft docking, and many other fields. Meanwhile, star sensors have also commercialized rapidly. Companies producing star sensors can be found not only in the U.S., but also in Germany, France, Belgium, the Netherlands, and many other countries. Among them, the U.S. has the largest number of such institutions, e.g., Ball Corporation, EMS Technologies Inc., Corning OCA Corporation, Jet Propulsion Laboratory (JPL), Lawrence Livermore National Laboratory (LLNL), Honeywell Technology Solutions Lab, etc. Some universities, such as Texas A&M University and the Technical University of Denmark, have also conducted in-depth research in star sensor technology. Figure 1.10 and Table 1.3 respectively demonstrate some typical CCD star sensors and the performance indicators of several typical CCD star sensors.

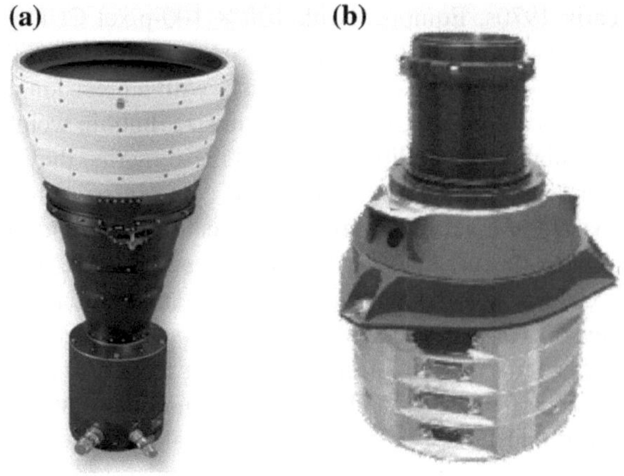

Fig. 1.10 Typical CCD star sensors **a** CT-601 star sensor developed by BALL Corporation, U.S. **b** ASTRO-15 star sensor developed by Jena-Optronik GmbH, Germany

Table 1.3 Performance indicators of several typical CCD star sensors

Indicator model	FOV (°)	Accuracy (″) (1σ)	Data update rate (Hz)	Weight (kg)	Consumption (W)
Corning OCA, U.S.	25 × 25	20	2	1.2	8.5
BALL CT-631, U.S.	20 × 20	12	5	2.5	8
Jena ASTRO-15, Germany	13.8 × 13.8	1	4	4.3	10

Mature as it now is, CCD technology has its inherent limitations in the fact that it is incompatible with deep submicron ultra-large-scale integration technology, which the photosensitive pixel array can only be realized on one chip, and that other functional units cannot be integrated on the same chip, complicating the imaging system and making it a multi-chip system. A typical star sensor based on a CCD imaging system weighs 1–7 kg and consumes 7–17 W of power. In addition, a CCD array requires a unique clock-driven pulse, various operating voltages, and a perfect charge transfer. The production process of CCD is complex and the cost is rather high. All these limitations of CCD make it hard to reduce the size, weight, and power consumption of a CCD-based imaging system.

Since the 1990s, people have set even higher demands for the weight, power consumption, and radiation resistance of star sensors. As a potential alternative to CCD technology, Active Pixel Sensor (APS) technology was developed for star sensors by the Jet Propulsion Laboratory (JPL) in the US. APS-based CMOS image sensors are superior to CCD image sensors in the following aspects:

(1) Easily integrated and equipped with simple interfaces. The photosensitive array, driving and control circuit, analog signal processor, A/D converter, all-digital interface, and other components are easily integrated onto one chip. This single-chip digital imaging system simplifies the electronic design of star sensors, decreases the number of peripheral circuits, and reduces the

size and weight of the imaging circuit system. The technology is thus favorable for the miniaturization of star sensors.

(2) Highly radiation resistant. The results of ground tests and space application suggest that the radiation resistance of CMOS image sensors significantly exceeds that of CCD technology.
(3) Low in power consumption. Through photoelectric conversion, CMOS image sensors can directly generate current signals. The only requirement is a single 5 V or 3.3 V power supply, and the consumption is 1/10 of that of a CCD image sensor.
(4) Flexible in data reading. Embedded in the pixel, the photodetection and output amplifier of CMOS image sensors can be separately located and read, just like DRAM.

Thanks to the above advantages, CMOS image sensors have been rapidly and widely adopted in star sensors. CMOS-based star sensors, often called third-generation star sensors, have been the focus of study in the field of star sensor technology over the past decade. Many institutions have poured huge human and material resources into relevant research. It is noteworthy that with the development of CMOS technology, the resolution and sensitivity of CMOS imaging devices has significantly improved in recent years. Figure 1.11 presents some typical CMOS star sensors, and Table 1.4 demonstrates the performance indicators of several typical CMOS star sensors.

1.3.3 Development Trends in Star Sensor Technique

Miniaturization, intellectualization, and low cost are future trends in the design of spacecraft. Correspondingly, the function, size, power consumption, and other

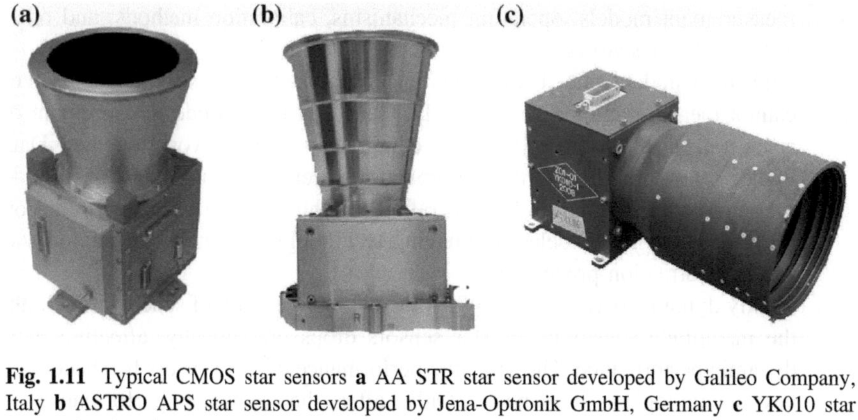

Fig. 1.11 Typical CMOS star sensors **a** AA STR star sensor developed by Galileo Company, Italy **b** ASTRO APS star sensor developed by Jena-Optronik GmbH, Germany **c** YK010 star sensor developed by Beijing University of Aeronautics and Astronautics, China

Table 1.4 Performance indicators of several typical CMOS star sensors

Indicator model	FOV (°)	Accuracy (″) (1σ)	Data update rate (Hz)	Weight (kg)	Consumption (W)
Beijing University of Aeronautics and Astronautics YK010, China	20	2	10	0.975	2.43
Jena ASTRO APS, Germany	20	2	10	1.8	4
Sodern HYDRA, France	20	3.3	10	2.2	3
Galileo AA STR, Italy	20	6	10	1.42	3.9
Aeroastro MST, U.S.	25	23.3	1	0.43	2

features of attitude-measuring instruments are faced with higher requirements. Currently, star sensor technology is developing rapidly, with trends in the following aspects:

(1) Autonomous navigation. The new generation star sensors have become integral attitude-measuring devices, with the capability of implementing star identification, star matching, and attitude calculation independently. Therefore, the ability to autonomously capture attitude without relying on external devices is an important indicator to measure the performance of star sensors and an inevitable trend in the development of star sensor technology as well.
(2) Higher accuracy in three-axis attitude measurement. Star sensors are the attitude measurement devices with the highest precision in spacecraft at present. However, modern space missions have set higher demands for the attitude control precision of spacecraft, calling for even more accurate attitude measurement. Restricted by the structure of conventional star sensors, the roll angle of star sensors is usually less accurate (around one magnitude lower than the pitch angle and yaw angle, usually larger than 10″). Nevertheless, space-based target tracking, high-resolution Earth observation, and other space missions require that three-axis attitude measurement should be more precise (≤2″). This means that in-depth research should be conducted to improve the measurement models, operating mechanisms, calibration methods, and other aspects of star sensors.
(3) High data update rate. Unlike inertial components, star sensors themselves cannot measure attitude continuously. Hence, in the attitude measurement of spacecraft, star sensors and inertial components usually work together. Data update rates should be further improved to meet the requirements of attitude measurement and control of spacecraft. The achievement of this goal relies on breakthroughs in photoelectric imaging technology and improvements in star image information processing.
(4) Highly dynamic performance. When the angular velocity of spacecraft is high, the magnitude sensitivity of star sensors drops dramatically, affecting their dynamic performance. The dynamic performance of star sensors has become a key technological bottleneck. Only after the dynamic performance is

significantly improved can star sensors be widely applied in maneuvering satellites, missiles, near space probes, and other situations.

(5) Miniaturization. Modern space missions, deep space exploration in particular, require that attitude sensors should be small in size, light in weight, and low in power consumption. These requirements also reflect the future development trends of star sensors. The application of CMOS imaging devices has laid a vital technological foundation for the miniaturization of star sensors.

(6) Reliability. In practical use, star sensors are faced with all kinds of interference, such as sunlight, earthshine, moonlight, etc. Moreover, the optical and electronic components of star sensors are subject to vacuum, temperature, radiation, and other space environment factors during long-term in-orbit operation. Therefore, the reliability of star sensors is an important indicator that directly relates to the performance of attitude and orbit control of spacecraft.

1.4 Introduction to Star Identification

Star identification is a vital prerequisite for the precise determination of the spatial attitude and position of a spacecraft. It identifies stars by matching the stars in the current FOV of a star sensor with reference stars in the guide star database. Generally, the angular distance of a star pair and the brightness of stars are considered as the basic characteristics of star images. The angular distance of a star pair, in particular, plays a crucial role in star identification. In this section, the fundamental principles and basic process of star identification are introduced first. Then, the performance of star identification is evaluated.

1.4.1 Principles of Star Identification

During the 1960s and 1970s, star sensors were widely applied in lunar orbiters, Apollo, Mariner, and other spacecraft. Astronauts took pictures of stars with film cameras and sent them back to the Earth for further processing. Hence, a large number of star images had to be manually analyzed and measured. In this context, Junkins came up with the idea of developing a universal star identification algorithm, which became the earliest star identification technology. With the rapid development of aerospace technology, the autonomous navigation function of spacecraft requires that star identification approaches should meet higher requirements in autonomy, speed, and precision.

For star sensors, star identification amounts to searching for guide stars in a star catalog (celestial coordinate system) corresponding to the measured stars in the star image [9], as shown in Fig. 1.12.

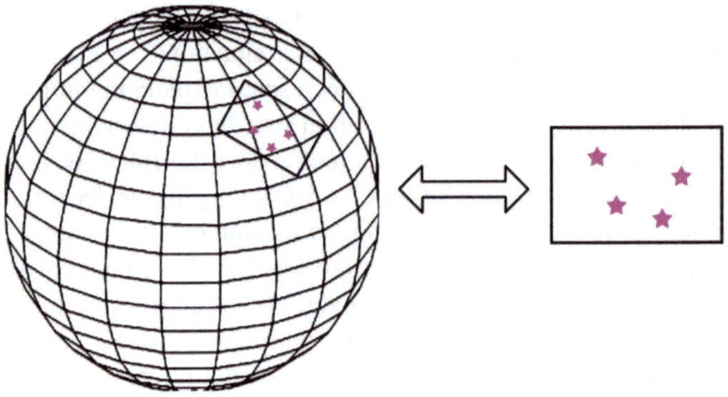

Fig. 1.12 Star identification

Generally speaking, a star sensor has at least two operating modes, i.e., an initial attitude establishment mode and a tracking mode. During the initial moment of operation or when faced with lost-in-space problems caused by a malfunction, a star sensor will enter into Initial Attitude Establishment Mode. With no prior attitude information, full-sky star identification is needed at this stage. Full-sky star identification usually takes a relatively long time and requires a high identification rate. Once the initial attitude is established, the star sensor will enter into Tracking Mode. Using the attitude information observed in the previous frames of images, a star sensor can predict and identify the position of stars in the current frame. Star identification in the Tracking Mode is faster and easier to operate.

1.4.2 The General Process of Star Identification

There are several processes involved in star identification, including image pre-processing, feature extraction, and matching recognition [14, 15]. For convenience of matching recognition, a guide database recording the features of guide stars extracted in advance should be established before identification begins. Figure 1.13 demonstrates the process of star identification.

(1) Image Pre-processing

Image pre-processing covers two sub-processes, noise removal and star centroid extraction. Affected by the celestial background, dark current from the photosensitive device and other factors, star images taken may contain some noise. This noise is usually removed with linear filtering, median filtering, morphological filtering, or other approaches. Then, the coordinates of a star point can be determined through a centroid extraction algorithm. The optical system of a star sensor is usually properly defocused so that star energy is distributed over a 3×3 to

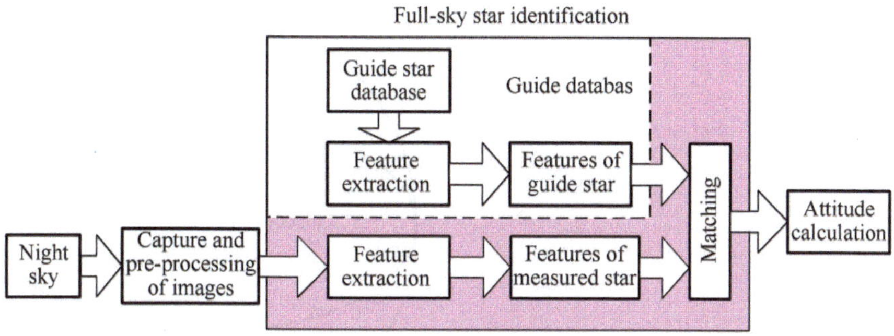

Fig. 1.13 Basic process of star identification

5 × 5 pixel area and the distribution of brightness is in accordance with a Point Spread Function. The locating result obtained this way is more accurate. According to the rules of star point energy distribution, there are some established centroid extraction algorithms, e.g., centroid method, square weighting centroid method, surface fitting, etc. With an appropriate centroid algorithm, the accuracy of locating may reach 1/20 pixel.

(2) Feature Extraction

There is relatively less available information for star identification than for common image recognition. Image recognition takes advantage of different types of information, such as gray level, contour, texture, etc. However, the available information for star identification is restricted to the coordinates and brightness of a star point. The former is described by the guide star's coordinates (right ascension and declination) with respect to the celestial coordinate system in the star catalog and the latter is measured by star magnitude. The brightness of a star (star magnitude) is usually considered unreliable, as differences in the spectrum response characteristics of image sensors may result in discrepancies between instrument magnitude and apparent magnitude. In addition, the magnitudes of some stars may change. Hence, star identification based on star magnitude should be avoided as much as possible.

In practical use, the angular distance between stars (i.e., the spherical angle between stars, usually called angular distance for short, as shown in Fig. 1.14) is considered as a feature useful for star matching. On one hand, the angular distance can be precisely calculated by observing star images. On the other hand, the angular distance between stars only changes minutely with time, and can be considered to vary within a very small threshold range. Furthermore, the angular distance remains unchanged in coordinate transformation and is a relatively reliable characteristic. For the reasons mentioned above, angular distance is considered a vital feature used in star identification and has been widely adopted in many types of identification algorithms.

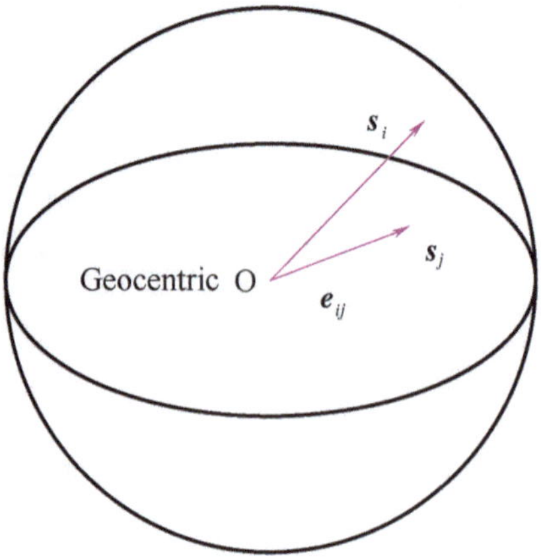

Fig. 1.14 Angular distance between stars

Generally speaking, most star identification algorithms require a more complicated feature extraction of the star information in star image. In this way, higher level features can be structured and the matching efficiency of a star identification algorithm can be improved.

(3) Construction of the Guide Database

Guide databases consist of two major parts: a Guide Star Catalog (GSC) and a feature database of guide stars. The GSC is a simple catalog compiled on the basis of the position (right ascension and declination) and brightness information of some guide stars from the basic star catalog. The brightness of these guide stars should be within a certain range. Except for constructing a GSC, star sensors also need to follow the feature extraction algorithm, establish the feature vectors of guide stars and store a guide star property database constructed on the basis of these feature vectors. The guide database should be complete and convenient to retrieve. Meanwhile, the quantity of data should be as small as possible in order to save storage space.

(4) Matching Recognition

Once the features of a measured star have been extracted following the method of constructing guide star characteristics, guide stars with similar features can be located in the guide database. If only one guide star is found to have characteristics similar to the measured star, then these two stars are considered as matched. The process of matching recognition and the approach involved in extracting features are closely related. After the initial matching results are obtained, follow-up matching is usually needed to verify the results. Generally, it takes the matching process a relatively long time to traverse the whole guide database. Fast search is thus a key technology in matching recognition. An efficient searching method can significantly improve the performance of a star identification algorithm. The results of matching recognition can be used for attitude calculation of star sensors or to provide reference information for subsequent matching.

1.4.3 Evaluation of Star Identification

Currently, there are various star identification algorithms. However, due to the differences in specific indicators and application background of star sensors, no unified and recognized evaluation standard has yet been established to assess the performance of these algorithms. Reviewing the present literature on star identification algorithms for star sensors, the evaluation and comparison are usually done in simulation conditions according to the following aspects [16]:

(1) Robustness

Robustness is mainly used to assess the impact of different kinds of interference on a star identification algorithm. Under the impact of a certain kind of interference, robustness is usually measured by the statistical results of the identification rate of an algorithm in repeated recognition tests with different boresight directions. The most frequently used types of interference are noise and interfering stars.

Corresponding to the two kinds of information used in star images, i.e., the position and brightness of star points, noise is grouped into two categories in simulation: star position noise and magnitude (brightness) noise. The positional deviation of star points is mainly caused by calibration errors in the star sensor (e.g., focal length measurement errors, lens distortion, optic axis offset errors, etc.) and star algorithm errors. Centroiding with sub-pixel-level accuracy can be obtained through high-precision calibration and fine locating of star points. Nonetheless, in order to inspect an algorithm's resistance to interference from position noise, a relatively large position noise is usually adopted to comprehensively investigate how it performs. Magnitude noise reflects the accuracy level of an image sensor's sensitivity toward stars' magnitudes. Though star magnitude is considered unreliable, common star identification algorithms still take advantage of it to a greater or lesser degree for faster and more accurate identification. In addition, the introduction of magnitude noise may increase or reduce the number of measured stars (stars with magnitudes that approach the observation limit of the star sensor) in the FOV. Hence, it is necessary to evaluate the impact of magnitude noise on star identification algorithms.

There are two types of interfering stars. The first type is so-called *unexpected stars*, including planets, nebulae and dust, space debris, etc. It is difficult to distinguish the imaging targets of artificial stars from those of common stars in a measured star image. Moreover, since image sensors have a limited capability of distinguishing star magnitude, dimmer stars are sometimes captured. However, there are no matching guide stars for these dimmer stars. Another type of interfering star is *missing stars*, i.e., stars that should have been captured but fail to show up in the observed FOV for some reason. Through simulation experiments, conditions with these two kinds of interfering stars are evaluated respectively and their influences on identification rate are analyzed.

(2) Identification Time

Full-sky star identification algorithms are used to determine the attitude of spacecraft in lost-in-space situations. Thus, the identification time should be as short as possible so that the system can rapidly establish initial attitude. For the above reason, the identification time of star identification algorithms is a key indicator in the design of star sensors. In simulations, the mean identification time of different identification algorithms are usually compared on the same hardware platform.

(3) Storage Capacity

In practical operation of star sensors, the star identification algorithm is usually run by an embedded CPU processor. The guide database is stored in ROM and loaded into RAM when the program is running. Since the storage capacities of ROM and RAM are limited, an algorithm's demand for storage capacity should be taken into account in the design of a star identification algorithm.

1.5 Star Identification Algorithms and Development Trends

Since star sensors came into being, scholars and researchers have put much effort into developing methods for star identification. At present, numerous star identification algorithms are available. In line with the methods of extracting features, these algorithms can be roughly grouped into two categories [17]:

(1) Subgraph isomorphism algorithms. This type of algorithm regards the angular distances between stars as sides, stars as vertexes, and the measured star image as a subgraph of the full-sky star image. A feature database is constructed in a certain manner, using angular distances directly or indirectly and regarding lines (angular distance), triangles, quadrangles, etc., as basic matching elements. Taking advantage of the combination of these basic matching elements, a corresponding match for the measured star image can be determined once the only area (subgraph) fitting the matching conditions is located in the full-sky star image. Conventional star identification algorithms, including polygon algorithms, triangle algorithms, group match algorithms, and others, all belong to the category of subgraph isomorphism algorithms. This type of algorithm is relatively mature and has been widely adopted.

(2) Star pattern recognition algorithms. This type of algorithm endows each star with a unique feature—a *star pattern*, which is usually represented by the geometric distribution features of other stars within a certain neighborhood. In this way, identifying stars becomes in essence searching for a guide star in the star catalog whose star pattern most resembles that of the measured star. Hence, this kind of algorithm is more like solving pattern identification problems. The most representative examples are grid algorithms.

This section introduces typical existing star identification algorithms and discusses their development trends.

1.5.1 Subgraph Isomorphism Algorithms

(1) Polygon Angular Matching Algorithms

Proposed by Gottlieb [18], polygon angular matching algorithms operate in the following way. Two measured stars are selected and their angular distance is calculated according to the equation below

$$d_m^{12} = \cos^{-1}(s_1 \cdot s_2).$$

Here, s_1 and s_2 stand for the direction vectors of the two stars in the star sensor coordinate system. This angular distance is then compared to all angular distances of star pairs stored in the guide database. Once two guide stars whose angular distance (i,j) satisfies

$$\left| d(i,j) - d_m^{12} \right| \leq \varepsilon \tag{1.8}$$

are found, (i,j) will be considered as a match for the two measured stars. Here, ε represents the error tolerance of angular distance. If more than one (i,j) satisfies the above conditions, a third star should be selected and similar comparison should be conducted to find a match for the angular distance between the third star and the previous ones. More stars are to be selected until only one match remains.

Polygon angular matching algorithms are relatively simple and realizable. However, when there are a large number of guide stars, the algorithms become complicated, requiring longer matching times and relatively large storage capacity. The increasing number of measured stars used for the matching (i.e., the increasing number of polygon sides) makes it more complex to determine the direction of angular distance. Hence, these algorithms usually require prior information of the pointing direction of the star sensor boresight.

(2) Triangle Algorithms

Triangle algorithms are the most frequently used and mature algorithms at present [9, 19, 20]. Similar to the fundamental principle of polygon angular matching algorithms, triangle algorithms utilize angular distances among three stars as their matching feature, as shown in Fig. 1.15. The matching triangle can be denoted as $(d_m^{12}, d_m^{23}, d_m^{13})$ or $(d_m^{12}, \theta, d_m^{13})$. Normally, $(d_m^{12}, d_m^{23}, d_m^{13})$ is stored in the navigation pattern database in accordance with the value of angular distances in ascending (or descending) order. The identification of triangle algorithms is in essence the search for the matching triangle which most resembles the observed triangle in the navigation pattern database.

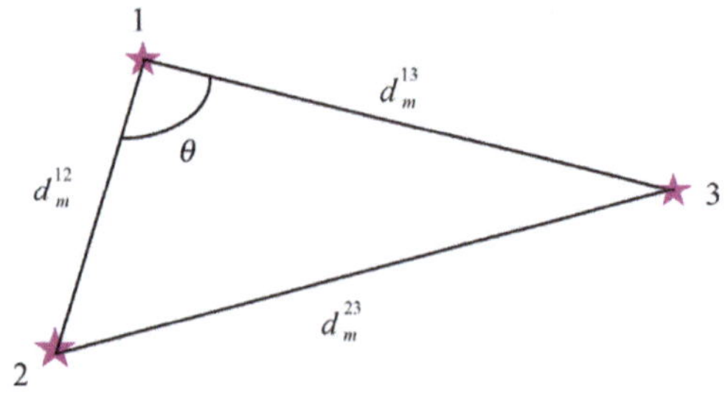

Fig. 1.15 Triangle algorithm

Since triangle algorithms store navigation triangles, the required storage capacity of guide databases is usually large. In addition, the selection of proper navigation triangles is vital for triangle algorithms. Quine, Heide, and other researchers [21] improved the current triangle selection rule in triangle algorithms, reducing the required storage capacity of guide databases remarkably and enhancing the identification rate. However, these approaches rely heavily on relatively precise star magnitude information. On the basis of triangle algorithms, Mortari proposed the Pyramid Algorithm [22, 23], which assesses the validity of the identification of a triangle algorithm by selecting a star outside the triangle. In this way, the possibility of redundant matches is reduced. Scholl introduced star magnitude information into triangle algorithms [24] and formed a six-feature vector. Though his method can also reduce the possibility of a redundant match, misidentification may occur due to inaccuracies in magnitude information.

(3) Group Match Algorithms

First put forward by Kosik [25] and further studied by Van Bezooijen and other researchers [26], group match algorithms work in the following way: A star is selected as the primary star (Pole Star, star marked as 1 in Fig. 1.16) from the measured star image (which contains at least four to five stars). Stars other than the primary star are called companion stars (Satellite Stars). Each companion star forms a star pair with the primary star, represented by d_m^{1n}. Similar to polygon angular matching, a matching star pair which meets the requirements is searched for among the guide stars. Denote the set of guide star pairs corresponding to d_m^{1n} as $\Re^{1n}$. The guide star that matches the primary star should be in the intersection of these sets $\left(\bigcap_{n=2}^{n=5} \Re^{1n}\right)$ and should be the guide star with the highest frequency of occurrence in $\Re^{1n}$ $(n = 2, \ldots, 5)$.

Group match algorithms organize feature patterns in the form of angular distance, and thus put a huge demand on guide database storage capacity. In addition, the identification rate is easily influenced by interfering stars. Through experiments, DeAntonio and other researchers carried out in-depth analysis of some deficiencies of Van Bezooijen's algorithm [27].

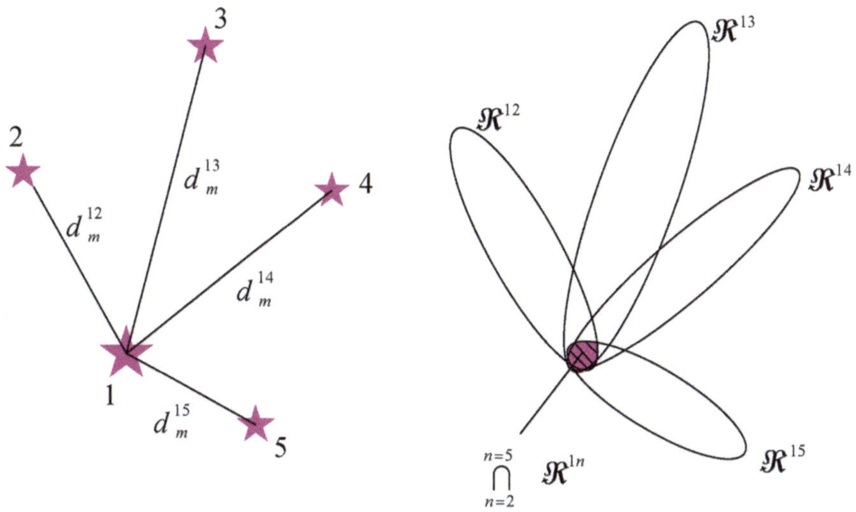

Fig. 1.16 Group match algorithm

1.5.2 Star Pattern Recognition Class Algorithms

(1) Grid Algorithms

First proposed by Padgett [17], grid algorithms demonstrate the geometric distribution features of companion stars in the neighborhood of the measured star with grids, and adopt them as the feature pattern of stars. The generation process of grid algorithms is illustrated in Fig. 1.17. A primary star and its pattern radius are determined first. Then, a *neighbor star* (i.e., the star nearest to the primary star outside a certain neighborhood radius) is identified. The line connecting the primary star and the neighbor star is considered as a coordinate axis, in accordance with which the FOV is rotated. The FOV is then divided into grids and the feature pattern of stars is constructed in the end. To accelerate the speed of matching, grid algorithms utilize a lookup table (LT) to store star features.

Compared with conventional algorithms, grid algorithms enjoy a relatively higher identification rate, a faster identification speed and allow a smaller guide database capacity. Nevertheless, they have one weak point. When an inappropriate neighbor star is selected, the wrong pattern generated may result in identification failure. Since it is not easy to select an appropriate neighbor star, the identification rate is affected to some extent. On the basis of grid algorithms, Clouse put forward a self-adaptive threshold grid algorithm based on Bayesian Theory [28]. The algorithm can further improve the accuracy of identification and is applicable in very small FOV (2°–3°) situations. The identification rate of grid algorithms is relatively low when there are only a few stars. Hence, grid algorithms usually require that there should be a large number of measured stars in the FOV.

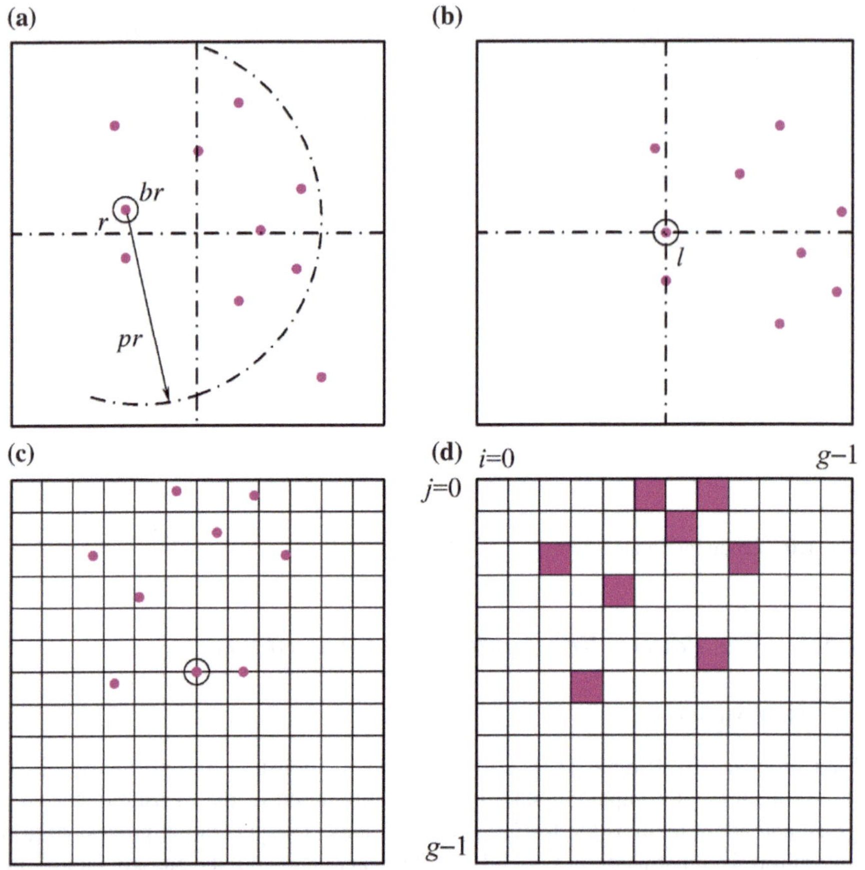

Fig. 1.17 Generation process of star pattern in grid algorithm **a** Determine the primary star *r* and its pattern radius *pr* **b** Shift the FOV and determine location star *l* **c** Rotate the FOV **d** Construct pattern

(2) Identification Algorithms Based on Statistical Features

Udomkesmalee put forward a star identification approach based on statistical features of stars [29]. In this approach, the statistical features of a companion star in the neighborhood is regarded as the star pattern, which is denoted as **x**

$$\mathbf{x} = (n, \mu_{\mathrm{m}}, \sigma_{\mathrm{m}}, \mu_{\mathrm{d}}, \sigma_{\mathrm{d}}, \sigma_{\theta}) \tag{1.9}$$

Here, n stands for the number of companion stars in the neighborhood. μ_{m} and σ_{m} represent the mean value and the variance of the magnitude (brightness) of the companion star, respectively. Similarly, μ_{d} and σ_{d} represent the mean value and the variance of the angular distance between the companion star and the primary star, respectively. The variance of the angle between neighboring companion stars is

Fig. 1.18 Star identification approach based on statistical features

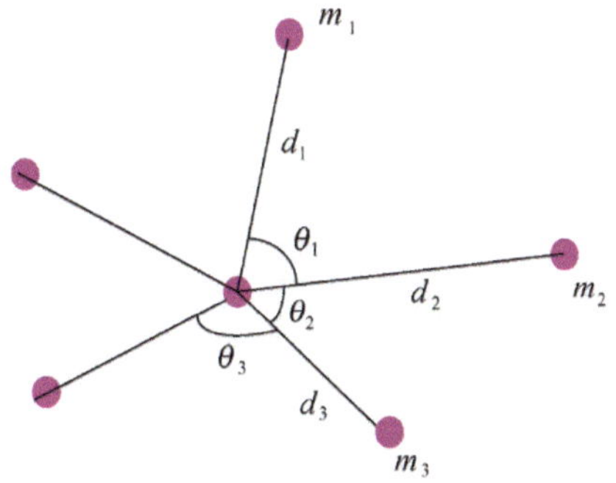

expressed as σ_θ, as shown in Fig. 1.18. The correctness of identification is verified through posterior probability of star image sequence acquired in different observed celestial areas. This algorithm is relatively complex and takes a longer time to identify stars. In addition, the construction of star patterns by this algorithm is subject to the influence of interfering stars. Hence, wrong star patterns of measured stars near the edge of the FOV are sometimes generated.

(3) Identification Algorithms Using a Star Axis Image Template

Kim proposed a star identification method which utilizes a star axis image template [30], i.e., organizing the information of stars around a single star pair into a star pair pattern. In the construction of the guide database, a guide star in a certain celestial area is first imaged onto a plane. The two brightest stars in the image are selected and considered as a star pair. Then, the image is rotated in the manner shown in Fig. 1.19. The template matrix of the star pair is calculated in accordance with the position of neighboring star points. For convenience of retrieval, angular distances of star pairs are stored in order.

For identification, the two brightest stars are first selected. The image is then rotated according to the positions of the two stars. In line with the angular distance of the two stars, candidate star pairs are selected by searching the catalog. The template matrix of the measured star pair is calculated, and the matching template

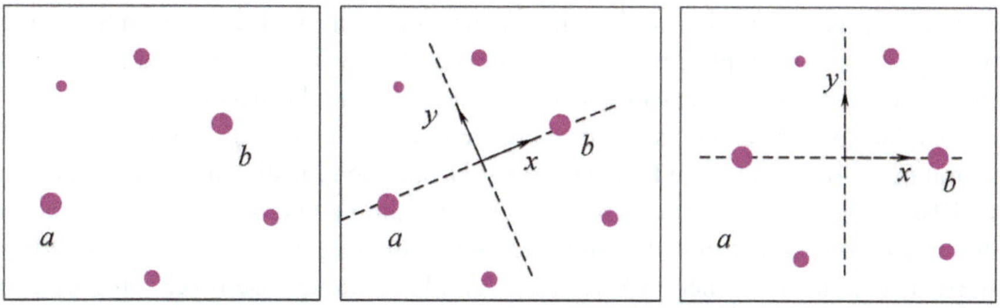

Fig. 1.19 Rotation of star image

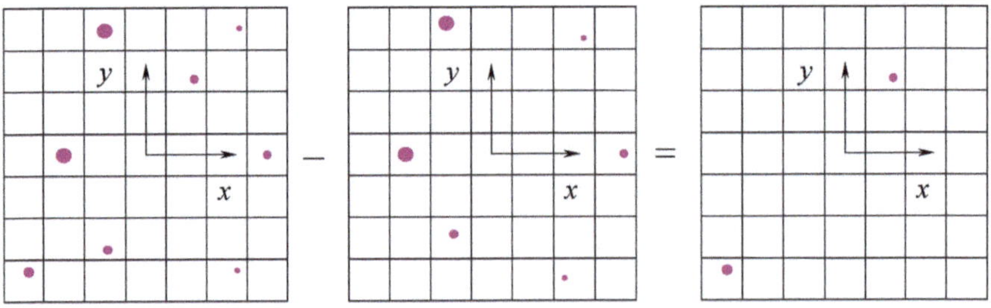

0	0	1	0	0	1	0
0	0	0	0	1	0	0
0	0	0	0	0	0	0
0	1	0	0	0	0	1
0	0	0	0	0	0	0
0	0	1	0	0	0	0
1	0	0	0	0	1	0

−

0	0	1	0	0	1	0
0	0	0	0	0	0	0
0	0	0	0	0	0	0
0	1	0	0	0	0	1
0	0	0	0	0	0	0
0	0	1	0	0	0	0
0	0	0	0	0	1	0

=

0	0	0	0	0	0	0
0	0	0	0	1	0	0
0	0	0	0	0	0	0
0	0	0	0	0	0	0
0	0	0	0	0	0	0
0	0	0	0	0	0	0
1	0	0	0	0	0	0

Fig. 1.20 Comparison and matching of templates

matrix for the guide star pair is searched for in the results of screened angular distances of star pairs. The templates are all organized in a discrete manner. The comparison of templates is shown in Fig. 1.20.

1.5.3 *Other Algorithms*

(1) Star Identification Algorithms Based on Neural Networks

With a unique ability of automatic learning and clustering, neural networks have been widely applied in star identification. Strictly speaking, neural networks are only used at the matching stage of star identification. Once the feature patterns of stars have been established, neural networks can be used for matching. Thanks to their strong generalization ability, neural networks can seek out the most similar prototype in an incomplete pattern. Hence, neural networks can be used to search in the guide stars for the measured star with the most similar pattern.

Scholars and researchers have conducted research on star identification with neural network models. Lindsey put forward a star identification algorithm based on an RBF network [31]. It quantizes and encodes the number of companion stars in the neighborhood distributed within the radius, treating it as the feature pattern of stars. Then, the RBF network is used for identification. Bardwell proposed an identification algorithm based on a Kohonen network [32]. Similar to grid algorithms, this approach selects a location star as the primary star. With the line connecting two stars as a baseline, the distribution features of companion stars are extracted. On the basis of grid algorithms, Li and other researchers took the features generated as feature vectors of a BP network for star identification [33]. Algorithms based on neural networks usually require more time since training is needed. As the number of guide stars grows, the number of star patterns and the scale of the network increase correspondingly. Therefore, it is hard for these algorithms to work in real-time and most of them are still only studied in simulation experiments.

(2) Star Identification Algorithms Based on Genetic Algorithms

Derived from biological evolution and population genetics, genetic algorithms (GA) have multiple merits. They are not restricted by the properties of functions and can realize global search and global convergence. Since GA came into being, they have been applied in many fields. The process of star identification by star sensors can be treated as a process of combinatorial optimization, during which GA can be used to search for the optimal combination. Paladugu [34] and Li [35] introduced GA into star identification and have obtained satisfactory results.

(3) Star Identification Algorithms Based on Hausdorff Distance

Hausdorff distance is used to evaluate the similarities between two images. No point-to-point correspondence of images is required to be established for calculating Hausdorff distance. Hence, it is suitable for application to images affected by noise or serious distortion. Star identification based on Hausdorff distance adopts the right ascension, declination and magnitude of stars in the basic star catalog as feature vectors. By calculating the Hausdorff distance between stars in the GSC and the star image, star identification can be accomplished [36]. This approach requires no prior knowledge and enjoys a relatively high identification rate. However, its disadvantage is that the identification rate of Hausdorff distance may be affected by its directional property, which makes the distance extremely sensitive to the rotation angle of the focal plane.

(4) Star Identification Algorithms Based on Singular Value Decomposition (SVD)

Generally, a matrix B can be obtained through the orthogonal transformation of matrix A, which is formed by a set of three-dimensional vectors. During the transformation, the three singular values of A and B remain the same. Taking advantage of this property, these three singular values can be considered as features and used in star identification [37]. For a particular frame of a measured star image, the four brightest stars are selected in the FOV and arranged in descending order according to their magnitudes. The matrix constituted by the corresponding vectors is decomposed and the three singular values are obtained. These singular values are taken as the feature vectors of the star image. By simply examining whether or not the three singular values are equal, the matching of stars can be accomplished. The advantage of this approach is that only three singular values are extracted as features in the end regardless of the number of vectors. During the process of Singular Value Decomposition (SVD), the optimal estimated attitude of a spacecraft can be obtained through simple calculation of singular vectors. The identification of stars is simplified by omitting the process of star matching, which means that there is no need to match each measured star with its corresponding guide star in a star image.

1.5.4 Development Trends of Star Identification Algorithms

Current trends in studies of star identification algorithms are mainly focused on the following aspects:

(1) Efficient star feature extraction methods. Conventional star identification algorithms, regarding angular distance and its derivative forms as features, are relatively simple. However, these methods have their inherent limitations, such as their requirement for large storage capacity, dissatisfactory performance in real-time identification, and generally low identification rate. Though neural networks, genetic algorithms, and other artificial intelligence approaches have been introduced into star identification, they can only influence the robustness and identification speed of algorithms to a certain degree. The key of star identification still lies in efficient methods for extracting star features.
(2) Fast matching algorithms. The development of modern spacecraft requires that attitude measurement should meet higher requirements in terms of speed. Faster star identification means that spacecraft can establish accurate and effective attitude as soon as possible. Hence, the rapidity of star identification algorithms is also a vital indicator in the design of star sensors. Another key technique for increasing the speed of star identification algorithms is the proper organization of GSC and the optimization of the matching process.
(3) Reliability. First, as a prerequisite for accurate attitude output, the identification rate of star identification algorithms should be high enough. Second, star identification algorithms should have a degree of robustness within the allowable range of measurement errors. Therefore, an excellent star identification algorithm should be fault tolerant so that star identification can be conducted properly even in poor conditions.
(4) Autonomy. The autonomy of star identification algorithms can be interpreted as their intelligence, which is a vital feature of the new generation of star sensors. This autonomy is displayed in the following aspects. First, star identification and three-axis attitude output can be accomplished independently without prior information or other auxiliary equipment. Second, star sensors can autonomously choose appropriate identification parameters so that optimal identification can be realized. Third, exceptional cases can be handled properly without losing attitude.

1.6 Introduction to the Book Chapters

Star identification covers multiple fields of studies, including astronomy, image processing, pattern identification, signal and data processing, computer science, and many others. This book summarizes the findings of the author's team's research in the area of star identification over more than ten years. There are seven chapters, covering basics in star identification, star catalog, and star image pre-processing, principles and processes of algorithms, and hardware implementation and performance testing.

Chapter 1 is a general introduction, covering basics in celestial navigation, with a discussion on star sensors and star identification, and reviews algorithms used in star identification and recent development trends in this field. Currently, many identification algorithms have been developed. However, due to the space constraints of this book, this chapter only introduces several representative algorithms. Readers interested in other algorithms may refer to relevant literature. Chapter 2 deals with the preliminary work in star identification, covering star catalogs, selection of guide stars, processing of double stars, star image simulation, star centroiding, and calibration of centroiding errors. Chapter 3 is a brief introduction to star identification by using triangle algorithms, with emphasis on two modified ones, namely, a modified triangle algorithm using angular distance matching and a modified triangle algorithm using *P* vectors. Chapter 4 first introduces grid algorithms. Then it focuses on star identification using star patterns, including star identification using radial and cyclic star patterns, star identification using Log-Polar transformations, and star identification without calibration parameters. Chapter 5 discusses basic principles of star identification using neural networks. Two methods are presented—star identification based on neural networks using features of star vector matrices and that using mixed features. Chapter 6 introduces the tracking mode of star sensors and focuses on rapid star tracking by using star matching between adjacent frames, with simulation results presented and analyzed. Chapter 7, by taking RISC CPUs as an example, deals with hardware implementation, as well as hardware-in-the-loop simulation testing and field experiments in star identification.

References

1. Gan G, Qiu Z (2000) Navigation and positioning. National Defence Industry Press, Beijing
2. Zhang S, Sun J (eds) (1992) Strap-down navigation system. National Defence Industry Press, Beijing
3. Inglis SJ (1979) Planets, stars, and galaxies. Science Press, Beijing
4. Roth GD (1985) Astronomy: a handbook. Science Press, Beijing
5. Shen C, Sun G (1987) Celestial navigation. National Defence Industry Press, Beijing
6. SAO Star Catalog, http://tdc-www.harvard.edu/software/catalogs/sao.html
7. Zhang G (2005) Machine vision. Science Press, Beijing

8. Fang J, Ning X, Tian Y (2006) Principles and methods of autonomous navigation of spacecraft. National Defence Industry Press, Beijing
9. Liebe CC (1995) Star trackers for attitude determination. IEEE Trans Aerosp Electron Syst 10 (6):10–16
10. Schmidt U (2005) ASTRO APS—the next generation Hi-Rel star tracker based on active pixel sensor technology. AIAA guidance, navigation, and control conference and exhibit, San Francisco, California, 15–18 Aug 2005. AIAA 2005-5925
11. Liebe CC, Dennison ED, Hancock B et al (1998) Active pixel sensor (APS) based star tracker. Aerospace conference proceedings (vol 1, pp 119–127). IEEE, Aspen, US. 21–28 March 1998
12. Liebe CC, Alkalai L, Domingo G et al (2002) Micro APS based star tracker. Aerospace conference proceedings (vol 5, pp 2285–2299). IEEE
13. Anderson DS (1991) Autonomous star sensing and pattern recognition for spacecraft attitude determination. Ph.D. Dissertation, Texas A&M University
14. Wei X (2004) A research on star identification methods and relevant technologies in star sensor (pp 1–14). Doctoral Thesis of Beijing University Aeronautics and Astronautics, Beijing
15. Yang J (2007) A research on star identification algorithm and RISC technology application (pp 1–17). Doctoral Thesis of Beijing University Aeronautics and Astronautics, Beijing
16. Padgett C, Kreutz-Delgado K, Udomkesmalee S (1997) Evaluation of star identification techniques. J Guid Control Dyn 22(2):259–267
17. Padgett C, Kreutz-Delgado K (1997) A grid algorithm for autonomous star identification. IEEE Trans Aerosp Electron Syst 33(1):202–213
18. Gottlieb DM (1978) In: Wertz JR (ed) Star identification techniques, spacecraft attitude determination and control (pp 257–266). The Netherlands
19. Birnbaum MM (1996) Spacecraft attitude control using star field trackers. Acta Astronaut 39 (9–12):763–773
20. Liebe CC (1992) Pattern recognition of star constellations for spacecraft applications. IEEE AES Mag 28(6):34–41
21. Quine BM, Durrant-Whyte HF (1996) Rapid star pattern identification. SPIE 2739:351–360
22. Mortari D, Junkins J, Samaan M (2001) Lost-in-space pyramid algorithm for robust star pattern recognition. 24th annual AAS guidance and control conference, AAS 01–004
23. Samaan M, Mortari D, Junkins J (2001) Recursive mode star identification algorithms. AAS/AIAA space flight mechanics meeting, AAS 01-149
24. Scholl M (2019) Star field identification algorithm—Performance verification using simulation star fields. SPIE 1993:275–290
25. Kosik J (1991) Star pattern identification aboard an inertially stabilized spacecraft. J Guid Control Dyn 14(2):230–235
26. Bezooijen RV (1989) Automated star pattern recognition. Ph.D. Dissertation, Stanford University
27. DeAntonio L, Udomkesmalee S, Alexander J et al (1949) Star-tracker based all-sky, autonomous attitude determination. SPIE 1993:204–215
28. Clouse D, Padgett C (2000) Small field-of-view star identification using Bayesian decision theory. IEEE Trans AES 36(2):773–783
29. Udomkesmalee S, Alexander J, Tolivar F (1994) Stochastic star identification. J Guid Control Dyn 17(6):1283–1286
30. Kim H (2002) Novel methods for spacecraft attitude estimation. Ph.D. Dissertation, Texas A&M University
31. Lindsey C, Lindblad T (1997) A method for star identification using neural networks. SPIE 3077:471–478
32. Bardwell G (1995) On-board artificial neural network multi-star identification system for 3-axis attitude determination. Acta Astronaut 35:753–761
33. Li Chunyan, Li Ke, Zhang Yunlong et al (2003) Star identification based on neural networks. Chin Sci Bull 48(9):892–895

34. Paladugu L, Schoen M, Williams BG (2003) Intelligent techniques for star-pattern recognition. Proceedings of ASME, IMECE2003-42274
35. Li L, Zhang F, Lin T (2000) An all-sky autonomous star map identification algorithm based on genetic algorithm. Opto-Electron Eng 27(5):15–18
36. Quan W, Wang G, Fang J (2006) Improved star map identification algorithm based on Hausdorff distance. J Beijing Univ Aeronaut Astronaut 32(1):8–12
37. Juang JN, Kim H, Junkins JL (2003) An efficient and robust singular value method for star pattern recognition and attitude determination. NASA Langley Research Center, NASA/TM-2003-212142

Chapter 2
Processing of Star Catalog and Star Image

Processing of star catalog and star image is the groundwork for star identification. The star sensor process establishes the three-axis attitude of spacecraft by observing and identifying stars. Thus, star information is indispensable. The star information used by the star sensor process mainly includes the positions (right ascension and declination coordinates) and brightness of stars. Star sensor's onboard memory can store the basic information of stars within a certain range of brightness. And this simplified catalog is generally called Guide Star Catalog (GSC). To accelerate the retrieval of guide stars, partition of the star catalog usually has to be done, which plays an important role in enhancing star identification and star tracking. At the early design stage of the star image processing and star identification algorithms, simulation approaches have to be taken to verify their correctness and to conduct performance evaluation. Therefore, star image simulation lays the foundation for simulation research of the star sensor. The measuring accuracy of the star vector by star sensor directly reflects the star sensor's performance in attitude establishment. Meanwhile, the measuring accuracy of the star vector is closely linked to star sensor's performance quality in star centroiding accuracy. Thus, it is necessary to conduct research on highly accurate centroiding algorithms that can be used by the star sensor technique.

This chapter first introduces the composition of GSC and partition methods of the star catalog. It also discusses guide star selection and double star processing. Then, it introduces star image simulation and star centroiding. After this, it discusses the calibration of centroiding error.

2.1 Star Catalog Partition

Star catalog partition plays an important role in star identification. It can accelerate the retrieval of guide stars in the star catalog, speed up full-sky star identification

G. Zhang, *Star Identification*, DOI 10.1007/978-3-662-53783-1_2

and star identification with established initial attitude. This section introduces GSC and catalog partition methods and presents star catalog partition with an inscribed cube method.

2.1.1 Guide Star Catalog

The number of stars in the star catalog has much to do with star magnitude. With the increase in star magnitude, the number of stars in the star catalog increases drastically. Through statistical analysis, an empirical equation regarding the relationship between the total number of stars distributed in the full sky and the change of star magnitude is obtained as follows [1]:

$$N = 6.57 \cdot \mathrm{e}^{1.08\,\mathrm{Mv}} \tag{2.1}$$

N stands for the total number of stars distributed in the full sky. Mv stands for star magnitude. Table 2.1 shows different star magnitudes and their corresponding total numbers of stars in Star Catalog SAO J2000.

To meet the demand of star identification by star sensor method, stars brighter than (or whose Mv is less than) certain magnitude are selected from the standard catalog and then used to build a smaller catalog (GSC) that is appropriate for star identification. Stars selected in the GSC are called guide stars. GSC contains the basic information of a guide star: right ascension, declination, and magnitude. The selection of magnitude is related to the star sensor's parameters. On the one hand, magnitude should be comparable to the limiting magnitude that can be detected by star sensor, that is, stars that can be observed by star sensor should be included in the GSC. Thus, magnitude should be equal to or slightly greater than the limiting magnitude that can be detected by star sensor, and the number of guide stars within the field of view must meet the needs of star identification. On the other hand, magnitude should be as small as possible on the premise that normal identification

Table 2.1 Different star magnitudes and their corresponding total numbers of stars

Magnitude	Total number of stars
3.0	155
3.5	260
4.0	480
4.5	871
5.0	1571
5.5	2859
6.0	5103
6.5	9040
7.0	15,935
7.5	26,584
8.0	46,172

can be achieved, which not only reduces the capacity of GSC, but also speeds up identification. Meanwhile, the probability of a redundant match drops with the decrease in the total number of guide stars. For example, if the limiting magnitude that can be detected by star sensor is 5.5 Mv, a total of 5103 stars whose brightness is greater than 6 Mv can be selected to make up a GSC.

2.1.2 Current Methods in Star Catalog Partition

How to retrieve guide stars rapidly must be taken into account in the process of building a GSC. The rapid retrieval of guide stars is of great importance in star identification, especially those in the tracking mode or with prior attitude information. If the arrangement of guide stars in a GSC is irregular, the entire GSC has to be traversed. Obviously, this kind of searching is rather inefficient. Therefore, the celestial area is usually divided into several sub-blocks.

Current methods in star catalog partition are listed below.

(1) Declination Zone Method

Through this method, the celestial sphere is divided into spherical zones (sub-blocks) by planes that are parallel to the equatorial plane. Each spherical zone has the same span of declination [2]. And guide stars in GSC can be directly retrieved by using a declination value. The problem with this method lies in the extremely uneven distribution of each sub-block. The number of guide stars in the sub-blocks near the equator is far greater than those near the celestial poles. This method does not make use of the information of right ascension, and the sub-blocks for retrieval contain a large number of redundant guide stars. Thus, its retrieval efficiency is rather low.

(2) Cone Method

Ju [3] divides the celestial sphere by using the cone method, as shown in Fig. 2.1. This method views the center of the celestial sphere as the vertex and uses 11,000 cones to divide the celestial sphere into regions that are exactly equivalent in size. When the angle (ψ) between the axes of neighboring cones is equal to 2.5° and the cone-apex angle (θ) is equal to 8.85° (the FOV is $10° \times 10°$), the stars included in the FOV by any boresight pointing are sure to be located within a certain cone. Through this method, possible matching stars that may correspond to measured stars in the FOV can be listed rapidly if the approximate boresight pointing of the star sensor is known beforehand. Since cones are overlapping, one measured star may be included in different sub-blocks. Thus, this partition method sets a high demand for storage space.

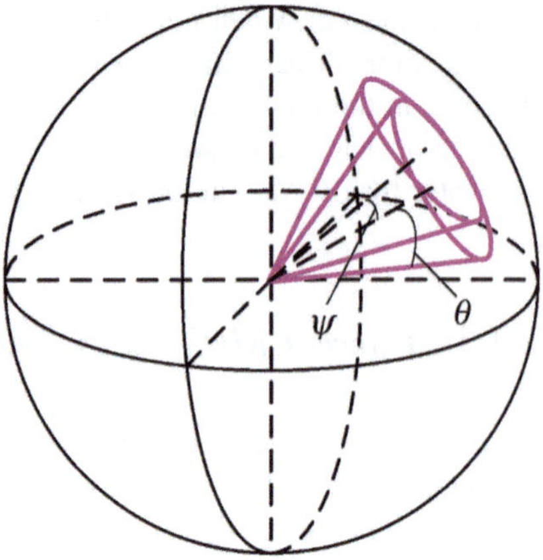

Fig. 2.1 Partition of the celestial area through cone method

(3) Sphere-Rectangle Method

Chen [4] uses the sphere-rectangle method to divide the celestial area. On the basis of a right ascension circle and declination circle, this method divides the celestial sphere into nonoverlapping regions, as shown in Fig. 2.2. The entire celestial area is divided into 800 sphere-rectangles and right ascension and declination are divided into 40 and 20 equal parts, respectively. Each sphere-rectangle stands for a span of 9° in the ascension direction and in the declination direction, respectively. It can be seen that this sphere-rectangle with right ascension and declination coordinates cannot be equal to the real FOV. The sizes of sphere-rectangles at different latitudes are different from each other, as shown in Fig. 2.2. In addition, the sub-blocks near the celestial poles cannot be completely represented by sphere-rectangles, so the retrieval of guide stars becomes consequently complicated. Since right ascension and declination coordinates are uneven themselves, the partition of the celestial sphere in that coordinate system cannot be even as well.

Fig. 2.2 Partition of the celestial area through sphere-rectangle method

2.1.3 *Star Catalog Partition with Inscribed Cube Methodology*

Zhang et al. [5, 6] use a completely different method. They divide the celestial area in the rectangular coordinate system and propose a star catalog partition with an inscribed cube method. This method realizes an even and nonoverlapping partition of the celestial area and the partition procedures are as follows.

① With inscribed cube, the celestial sphere is divided evenly into six regions, as shown in Fig. 2.3a. A cone is formed when the center of the celestial sphere is connected to the four vertices of each cube side, respectively. The cone is intersected with the celestial sphere and divides the sphere into six parts: S_1–S_6. The direction vectors of the central axis (v) of S_1 and its four boundary points are as follows:

$$v = (0,0,1)\, w_1 = (1,1,1)/\sqrt{3}\, w_2 = (1,-1,1)/\sqrt{3}\, w_3 = (-1,-1,1)/\sqrt{3}\, w_4 = (-1,1,1)/\sqrt{3} \tag{2.2}$$

And the direction vectors of the central axes (v) of S_2–S_6 and their four boundary points can be analogized and so on.

② Each part of S_1–S_6 can be further divided into $N \times N$ sub-blocks, as shown in Fig. 2.3b, c. In this way, the entire celestial sphere is divided into $6 \times N \times N$ sub-blocks. Besides, all sub-clocks are equivalent in size, the FOV

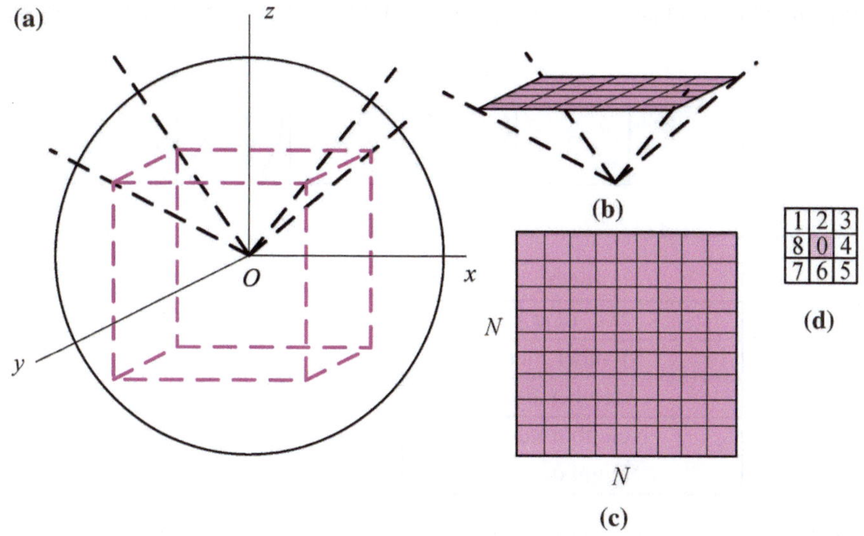

Fig. 2.3 Partition of the celestial area

being covered amounting to $90/N \times 90/N$.The direction vectors of each sub-block's central axis and boundary points can be acquired based on the direction vectors of S_1–S_6.

Based on this method, the celestial sphere is divided. The GSC is scanned and each guide star finds its corresponding sub-block. After this, a partition table is created, as shown in Fig. 2.4.

The partition table has $6 \times N \times N$ parts, each representing a sub-block. Each part records the following information:

Index: index number of a sub-block
(x, y, z): direction vector of the central axis of a sub-block
Member num: the total number of stars in a sub-block
Member list: list of stars in a sub-block
Neighbor num: number of neighboring sub-blocks (as shown in Fig. 2.3d)
Neighbor list: list of neighboring sub-blocks.

(x, y, z) is arranged by its magnitude in ascending (or descending) order. If the direction vector of the boresight pointing (or right ascension and declination coordinates) is known beforehand, the corresponding sub-block and its neighboring sub-block can be found quickly in the celestial area. The index number of the sub-block to which a guide star is affiliated is also stored in the GSC in order to retrieve its neighboring guide star rapidly from its index number. Thus, the GSC contains the direction vector and magnitude of a guide star and the index number of the sub-block to which it is affiliated. The GSC and partition table created in that way can realize the rapid retrieval from initial attitude (boresight pointing) or guide star index number to guide stars in a given neighborhood.

Take the FOV of 10° × 10°, for example. In order to make any sub-block and its neighboring sub-blocks (such as the 3 × 3 sub-blocks in Fig. 2.3d) incorporate the

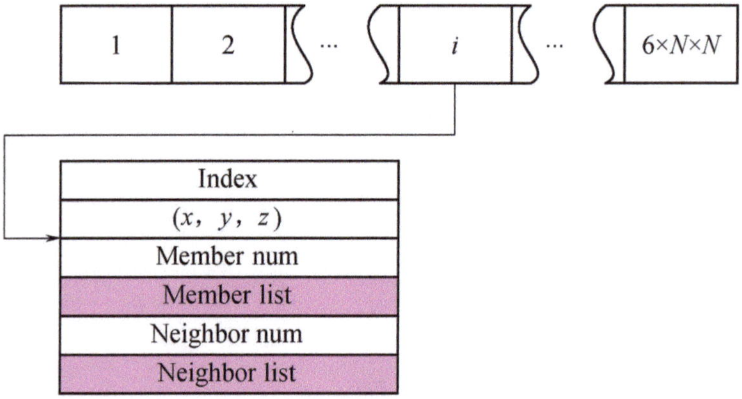

Fig. 2.4 Structure of the partition table

FOV as completely as possible, take $N = 9$, that is, divide the celestial area into $6 \times 9 \times 9 = 486$ sub-blocks and the size of each sub-block is $10° \times 10°$. Then there is no need to traverse the entire GSC to retrieve guide stars and the average search scope is only 9/486 = 1/54 that of before. After partition, the statistical results of the number of guide stars distributed in each sub-block are as follows:

the maximum number is 39;
the minimum number is 2;
the average number is 10.61.

2.2 Guide Star Selection and Double Star Processing

Guide star selecting is aimed at cutting down the number of guide stars as much as possible on the premise that correct star identification is guaranteed, which not only reduces the storage capacity required in star identification algorithms, but also speeds up star identification. It is thus an important work for enhancing star identification and reducing the capacity of the navigation database. Meanwhile, double star processing has implications for star identification. This section introduces guide star selecting and discusses the processing methods of a double star.

2.2.1 *Guide Star Selection*

Assume that guide stars are distributed evenly and randomly in the celestial area, the number of stars in the FOV approximately follows a poisson distribution [7]:

$$p(X = k) = \frac{\mu^k e^{-\mu}}{k!} \tag{2.3}$$

μ can be computed as follows: the spherical area of an FOV of $12° \times 12°$ is 0.04376 and the number of FOV accommodated by the entire celestial sphere is $4\pi/0.04376 = 287.02$. Thus, the average number of stars whose brightness is greater than 6 Mv in each FOV is 17.78 (μ = 5103/287.02 = 17.78).

Based on Poisson distribution, the probability of the number of stars in the FOV is shown in Table 2.2.

Table 2.2 Probability of the number of stars in the FOV based on Poisson distribution (FOV of $12° \times 12°$, Magnitude ≥ 6)

	$X \le 2$	$X \le 5$	$X \le 10$	$X \ge 20$	$X \ge 30$	$X \ge 40$	$X \ge 50$
Probability P (%)	0	0.04	3.38	32.97	0.50	0	0

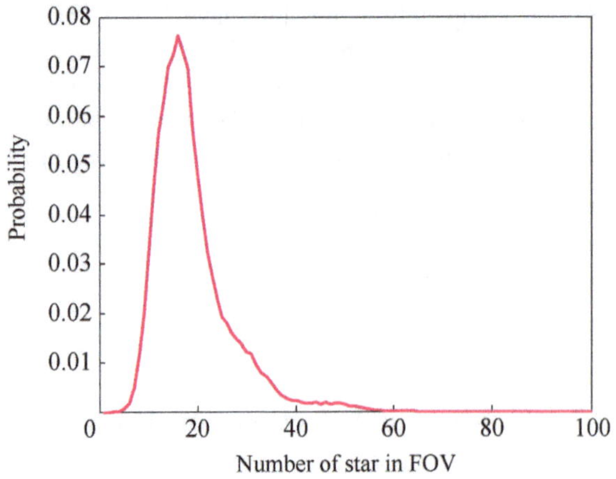

Fig. 2.5 Probability statistics of the distribution of guide stars in the FOV (FOV of 12° × 12°, Magnitude ≥ 6)

The statistical result (as shown in Fig. 2.5) shows the numbers of stars brighter than 6 Mv in an FOV of 12° × 12° by 100,000 random boresight pointings. The maximum number of stars in the FOV is 63, the minimum number is 2, and the average number is 17.8. The results computed by Poisson distribution are basically the same as the simulation results, as shown in Fig. 2.5. In fact, the distribution of stars in the celestial sphere is not even or random: the distribution near the celestial pole is sparse while that near the equator is relatively dense.

As for star identification, the number of measured stars in the FOV cannot be too small and must meet the minimum requirements of identification (≥3). If the number is too small, then the information that can be used would be relatively little and, thus, it would be difficult for identification. Meanwhile, accuracy e in attitude establishment is related to the number n of stars that are involved in attitude establishment in the FOV:

$$e = e_0/\sqrt{n} \tag{2.4}$$

Here, e_0 is the accuracy in attitude establishment of one star. Theoretically, the larger n is, the higher the accuracy in attitude establishment would be. Thus, in order to guarantee high accuracy in attitude establishment, the average number of stars in the FOV should not be too small. Generally, $n \geq 5$–6.

Correspondingly, the number of guide stars in each FOV should meet the needs of normal identification and attitude establishment. It can be noted that the number of stars in the FOV is relatively larger in some celestial areas where stars are densely distributed. Generally, excessive stars do not affect the results of identification and contribute little to the enhancement of accuracy in attitude establishment. In contrast, excessive stars mean more time in identification and a larger capacity of GSC and pattern database. In addition, excessive stars lead to too many redundant matches. Thus, the number of guide stars in the FOV should be as small as possible on the premise that normal identification and accuracy in attitude establishment can be guaranteed, that is, redundant guide stars should be eliminated in dense celestial areas in order to guarantee that the distribution of guide stars in the entire celestial area is as even as possible.

Based on the single magnitude threshold, there are too many stars in the FOV as a result of some boresight pointings, while there are no stars in the FOV of some celestial areas. Obviously, this method cannot guarantee the even distribution of selected guide stars in the celestial area. Based on the process of partition of GSC, guide star selection can be realized by traversing the distribution of guide stars in the full sky.

As described in the last section, the celestial area is divided into $6 \times 9 \times 9 = 486$ sub-blocks and then the direction vector of each sub-block's corresponding central axis can be obtained. 486 boresight pointings (direction vector) evenly distributed in the full sky resulting from this method and the angle between each pair of neighboring boresight pointings is 10°. Similarly, the full sky can be divided into $6 \times 100 \times 100 = 60{,}000$ boresight pointings that are evenly distributed, and the angle between each pair of neighboring boresight pointings is 0.9°. The 60,000 boresight pointings are scanned, and then guide stars located within the FOV (such as a circular FOV) by each boresight pointing are established.

If the number of guide stars in the FOV is lower than or equal to a certain threshold number C, then no processing is needed. Otherwise, it is necessary to arrange guide stars in the order of brightness (magnitude), keeping the brightest C stars and eliminating darker ones. The selection of C is related to the requirements put forth by star identification for the minimum number of guide stars in the FOV and accuracy in attitude establishment.

This method mainly takes into account the characteristic that brighter stars are easier to be sensitized by the star sensor method. Thus, it is reasonable to view brighter stars as guide stars. In the relatively dense celestial areas, redundant darker stars can be eliminated from GSC, while in the relatively sparse celestial areas, as many as possible guide stars should be retained.

Take $C = 6$, and Fig. 2.6 shows the probability statistics of the number of guide stars in the FOV after selecting. The minimum number of guide stars in the FOV is 2, the maximum number is 28, and the average number is 11.9. After selection, the total number of guide stars drops from 5103 to 3360. Compared with Fig. 2.5, Fig. 2.6 shows that the distribution of guide stars is more reasonably even than that before selection.

2.2.2 *Double Star Processing*

"Double star" is a different concept from that of "binary star" in astronomy. Astronomically, binary stars are two stars that are close to and revolve around each other. By contrast, here 'double star' means two stars that seem to be close in the direction of the line of sight (actually they may be far away from each other), and

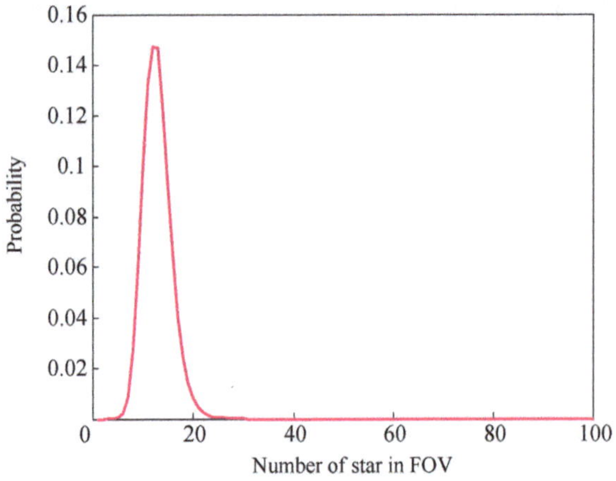

Fig. 2.6 Probability statistics of the distribution of guide stars in the FOV after selecting (FOV 12° × 12°, Magnitude ≥ 6)

the star spots formed by the two on the image plane of the star sensor cannot be separated from each other. The size of a star spot is related to the point spread function (PSF) of the optical system and its visual magnitude.

Generally, in order to improve the accuracy of the star position, a defocusing technique is often used to make the size of the image point range from 3 × 3 to 5 × 5 pixels. Assume the radius of PSF is one pixel, Fig. 2.7a, b show the gray distribution of the double star's spot images. Star spot images can be approximately represented by the Gaussian function:

$$f(x,y) = A\exp\left(-\frac{(x-x_0)^2+(y-y_0)^2}{2\sigma^2}\right) \tag{2.5}$$

A stands for the brightness of stars which is related to magnitude. Assume the binary threshold in the process of star spot extraction is T. The minimum distance between two stars to constitute double star must be d, as shown in Fig. 2.7c.

$$T = 2A\exp\left(-\frac{(d/2)^2}{2\sigma^2}\right) \tag{2.6}$$

Take $T = 80$, $\sigma = 1$ and $A = 255$. By computation, $d \approx 4$. Thus, two stars whose star spot positions are smaller than four pixels apart in the image plane are viewed as a double star. This means that two guide stars between which the angular distance is smaller than 0.047° (e.g., 12° × 12° in FOV, 1024 × 1024 in resolution) are treated as a double star.

A double star interferes in the process of star identification and, thus, general star identification algorithms cannot find correct matches for a double star. Meanwhile, a double star affects the identification of other stars. The traditional way of

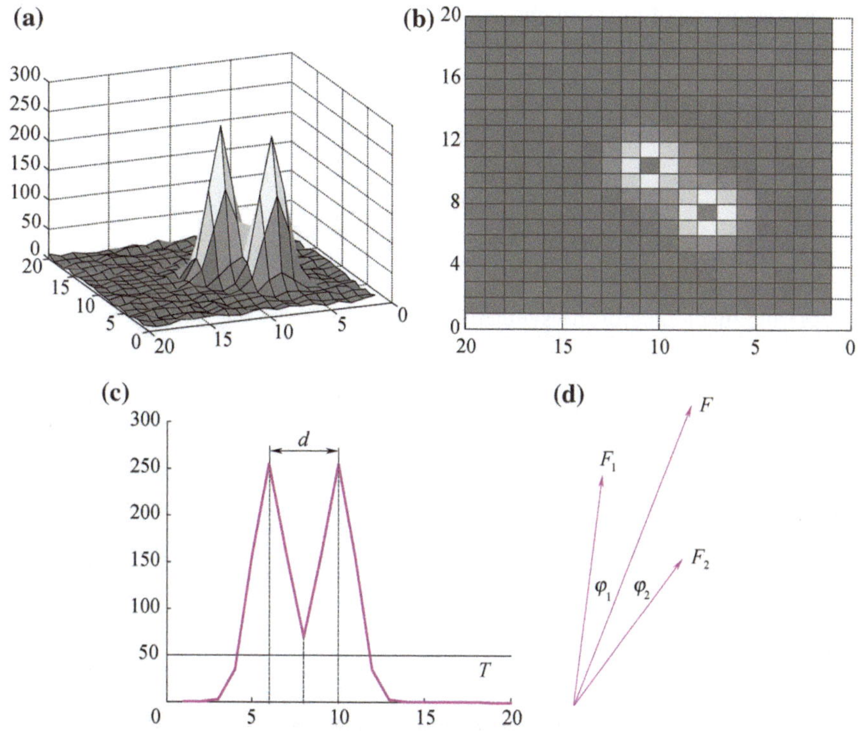

Fig. 2.7 Double star and double star processing

processing a double star is to eliminate them directly, which is feasible when the number of stars in the FOV is quite large. When the number of stars in the FOV is relatively small, eliminating the double star directly means throwing away some available information that is necessary for star identification. Considering that, the double star can be treated as a new "synthetic star" whose magnitude and orientation are synthesized by those of the double star. In fact, star spot images of the double star acquired by star sensor can be viewed as synthesized by the star spot images of the two stars.

Assume the magnitude of each of the double star is m_1 and m_2, respectively, and the direction vector of each is v_1 and v_2, as shown in Fig. 2.7d. Star's brightness is represented by the density of optical flow, and the brightness ratio of the two stars is:

$$\frac{F_1}{F_2} = \mathrm{e}^{(m_2 - m_1)/2.5} \tag{2.7}$$

The brightness of the synthetic star can be viewed as the synthesis of that of the two stars that constitute the double star, that is,

$$F = F_1 + F_2 \tag{2.8}$$

Thus,

$$\frac{F}{F_2} = \frac{F_1 + F_2}{F_2} = \mathrm{e}^{(m_2 - m)/2.5} \tag{2.9}$$

And the magnitude of the synthetic star is:

$$m = m_2 - 2.5\,\log\left(1 + \frac{F_1}{F_2}\right) = m_2 - 2.5\,\log\left(1 + \mathrm{e}^{(m_2 - m_1)/2.5}\right) \tag{2.10}$$

Assume the angular distance between the synthetic star and each of the double stars is φ_1 and φ_2, respectively, and the angular distance between the two stars is φ.

$$F_1\varphi_1 = F_2\varphi_2$$

$$\varphi = \varphi_1 + \varphi_2 = \varphi_1\left(1 + \frac{F_1}{F_2}\right) = \varphi_1\left(1 + \mathrm{e}^{(m_2 - m_1)/2.5}\right) \tag{2.11}$$

Thus, φ_1 and φ_2 can be computed, and the direction vector (v) of the synthetic star can be obtained:

$$v\,\sin\varphi = v_1\,\sin\varphi_1 + v_2\,\sin\varphi_2 \cong v_1\varphi_1 + v_2\varphi_2$$

$$v = (v_1\varphi_1 + v_2\varphi_2)/\sin\varphi \tag{2.12}$$

2.3 Star Image Simulation

There are mainly three approaches in star identification algorithm research: digital simulation, hardware-in-the-loop simulation, and field test of star observation. These three approaches correspond to three different stages in designing a star sensor and a star identification algorithm. At the early design stage, preliminary performance evaluation of the algorithm was done using digital simulation to determine appropriate identification parameters. Digital simulation is computer-based and is involved in the whole process of star image simulation, star image processing, and star identification. After the design of the star sensor finishes, star identification algorithms can be verified using the method of hardware-in-the-loop simulation.

Through hardware-in-the-loop simulation, star field simulator (SFS) generates star images. Then the imaging, processing, and identification of those generated star images are done by star sensor. Field tests of star observation are done during the night. Star images are photographed and then identified by the star sensor method in order to further verify the star identification algorithms. As for star identification algorithms, star image simulation is the fundamental work of the research. That is, when star sensor's attitude or boresight pointing is given, those star images photographed by star sensor can be simulated. Star image simulation mainly includes two processes: the imaging of star sensor and the synthesis of digital star images.

2.3.1 The Image Model of the Star Sensor

In the transformation of coordinates in the process of star image simulation, it is necessary to involve the celestial coordinate system, the star sensor coordinate system and the star sensor image coordinate system. The simple definitions of these coordinate systems are as follows [5]:

① Celestial coordinate system: With the celestial equator as its fundamental circle, the hour circle passing through the vernal equinox as its primary circle and the vernal equinox as its principal point, as shown in Fig. 1.2, this system uses right ascension and declination as its coordinates.

② Star sensor coordinate system: this is a system with the projection center o (the point on the boresight, with its distance from the focal plane as f, is called the optical center) as the origin of coordinates, boresight as z axis, and two straight lines that pass through point o and are parallel to both sides of the image sensor as the x and y axes. Figure 2.8 shows the front projection imaging of star sensor.

③ Image coordinate system: this system is a plane coordinate system of x and y axes that are parallel to both sides of the image sensor, with the center (principal point) of image sensor as the origin of the coordinates, as shown in Fig. 2.8.

Assume the coordinate of a star in the celestial coordinate system is (α_i, δ_i), the direction vector in the star sensor coordinate system is (x_i, y_i, z_i), and the coordinate of the imaging point in the image coordinate system is (X_i, Y_i).

The imaging process of star sensor consists of two steps: the rotation transformation of stars from the celestial coordinate system to the star sensor coordinate system and the perspective projection transformation from the star sensor coordinate system to the image coordinate system [5].

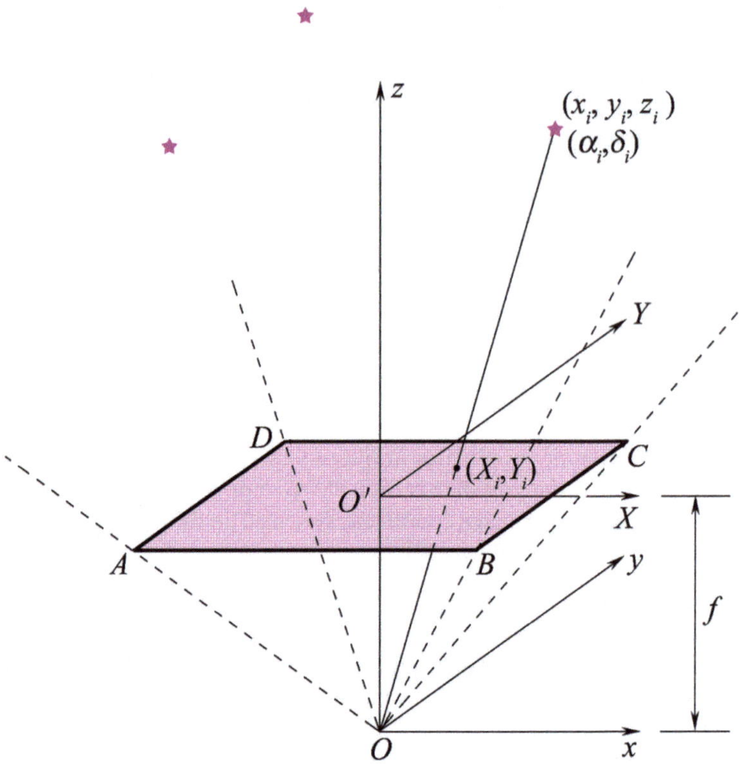

Fig. 2.8 Illustration of the star sensor coordinate system, the image coordinate system, and front projection imaging

(1) Rotation Transformation

Assume the attitude angle of the star sensor is $(\alpha_0, \delta_0, \phi_0)$. Here, α_0 stands for right ascension, δ_0 for declination, and ϕ_0 for roll angle. The rotation matrix (M) from the star sensor coordinate system to the celestial coordinate system can be expressed as:

$$M = \begin{pmatrix} \cos(\alpha_0 - \pi/2) & -\sin(\alpha_0 - \pi/2) & 0 \\ \sin(\alpha_0 - \pi/2) & \cos(\alpha_0 - \pi/2) & 0 \\ 0 & 0 & 1 \end{pmatrix} \begin{pmatrix} 1 & 0 & 0 \\ 0 & \cos(\delta_0 + \pi/2) & -\sin(\delta_0 + \pi/2) \\ 0 & \sin(\delta_0 + \pi/2) & \cos(\delta_0 + \pi/2) \end{pmatrix}$$
$$\begin{pmatrix} \cos\phi_0 & -\sin\phi_0 & 0 \\ \sin\phi_0 & \cos\phi_0 & 0 \\ 0 & 0 & 1 \end{pmatrix} = \begin{pmatrix} a_1 & a_2 & a_3 \\ b_1 & b_2 & b_3 \\ c_1 & c_2 & c_3 \end{pmatrix} \tag{2.13}$$

In this matrix,

$$\begin{aligned}
a_1 &= \sin\alpha_0\cos\varphi_0 - \cos\alpha_0\sin\delta_0\sin\varphi_0 \\
a_2 &= -\sin\alpha_0\sin\varphi_0 - \cos\alpha_0\sin\delta_0\cos\varphi_0 \\
a_3 &= -\cos\alpha_0\cos\delta_0 \\
b_1 &= -\cos\alpha_0\cos\varphi_0 - \sin\alpha_0\sin\delta_0\sin\varphi_0 \\
b_2 &= \cos\alpha_0\sin\varphi_0 - \sin\alpha_0\sin\delta_0\cos\varphi_0 \\
b_3 &= -\sin\alpha_0\cos\delta_0 \\
c_1 &= \cos\alpha_0\sin\varphi_0 \\
c_2 &= \cos\alpha_0\cos\varphi_0 \\
c_3 &= -\sin\delta_0
\end{aligned}$$

Since M is an orthogonal matrix, the rotation matrix from the celestial coordinate system to the star sensor coordinate system can be expressed as $M^{-1} = M^{\mathrm{T}}$. First, search image able stars with a circular FOV, and the right ascension and declination coordinates (α, δ) of stars that can be imaged on the image sensor should satisfy the following conditions:

$$\begin{aligned}
&\alpha \in (\alpha_0 - R/\cos\delta_0, \alpha_0 + R/\cos\delta_0) \\
&\delta \in (\delta_0 - R, \delta_0 + R)
\end{aligned} \tag{2.14}$$

Here, R stands for the radius of a circular FOV (R is half of the diagonal angular distance of the FOV, for example, the value of R in an FOV of 12° × 12° is equal to $6\sqrt{2}$ ($R = 6\sqrt{2}$). (α_0, δ_0) stands for the boresight pointing of star sensor. The direction vector of stars that satisfy Eq. (2.13) in the star sensor coordinate system can be expressed as:

$$\begin{pmatrix} x_i \\ y_i \\ z_i \end{pmatrix} = M^{\mathrm{T}} \begin{pmatrix} \bar{x}_i \\ \bar{y}_i \\ \bar{z}_i \end{pmatrix} \tag{2.15}$$

Here, $\begin{pmatrix} \bar{x}_i \\ \bar{y}_i \\ \bar{z}_i \end{pmatrix} = \begin{pmatrix} \cos\alpha_i\cos\delta_i \\ \sin\alpha_i\cos\delta_i \\ \sin\delta_i \end{pmatrix}$ is the direction vector of stars in the celestial coordinate system.

(2) Perspective Projection Transformation

The imaging process of stars on the image sensor can be represented by the perspective projection transformation, as shown in Fig. 2.8. After perspective projection, the coordinates of stars' imaging points are as follows:

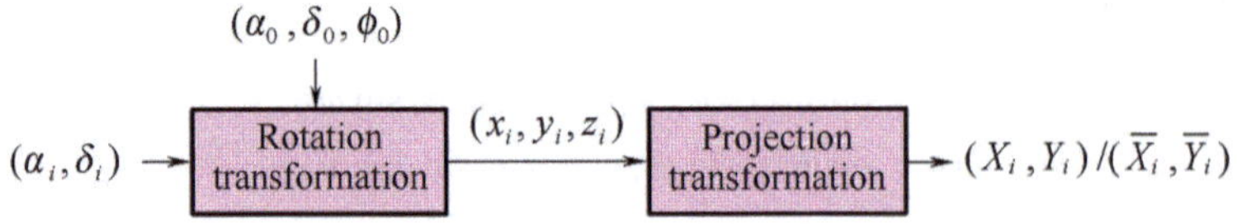

Fig. 2.9 Imaging process of star sensor

$$\begin{aligned} X_i &= f\frac{x_i}{z_i} = f\frac{a_1\bar{x}_i + b_1\bar{y}_i + c_1\bar{z}_i}{a_3\bar{x}_i + b_3\bar{y}_i + c_3\bar{z}_i} \\ Y_i &= f\frac{y_i}{z_i} = f\frac{a_2\bar{x}_i + b_2\bar{y}_i + c_2\bar{z}_i}{a_3\bar{x}_i + b_3\bar{y}_i + c_3\bar{z}_i} \end{aligned} \tag{2.16}$$

To sum up, the imaging process of star sensor can be illustrated by Fig. 2.9. The imaging model of star sensor can be expressed as follows:

$$s\begin{bmatrix} X \\ Y \\ 1 \end{bmatrix} = \begin{bmatrix} f & 0 & u_0 \\ 0 & f & v_0 \\ 0 & 0 & 1 \end{bmatrix} \cdot \begin{bmatrix} r_1 & r_2 & r_3 \\ r_4 & r_5 & r_6 \\ r_7 & r_8 & r_9 \end{bmatrix} \cdot \begin{bmatrix} \cos\beta\cos\alpha \\ \cos\beta\sin\alpha \\ \sin\beta \end{bmatrix} \tag{2.17}$$

In the above equation, s stands for nonzero scale factor, f for focal length of the optical system, (u_0, v_0) for optical center (the coordinates of the principal point), $(r_1–r_9)$ for transformation matrix from the celestial coordinate system to the star sensor coordinate system, and (α, β) for the right ascension and declination coordinates of the starlight vector in the celestial coordinate system. Equation (2.17) shows that the position coordinates (α, β) of stars in the celestial coordinate system (world coordinate system) are in a one-to-one correspondence with the positions (X, Y) of image points on the image plane of star sensor.

(3) Nonlinear Model

In fact, an actual optical lens cannot achieve perfect perspective imaging, but has different degrees of distortion. As a result, the image of the spatial point is not located at the position (X, Y) as described by the linear model, but within the actual image plane coordinates (X', Y') migrated due to the distortion of the optical lens.

$$\begin{cases} X' = X + \mathrm{d}x \\ Y' = Y + \mathrm{d}y \end{cases} \tag{2.18}$$

Here, dx and dy stand for distortion values which are related to the position of the star spot's coordinates in the image. Generally, there exist radical and tangential distortions in an optical lens. As for three-order radical distortion and two-order tangential distortion, distortions in the directions of x and y can be expressed as follows [8]:

$$\begin{cases} \mathrm{d}x = \bar{x}\left(q_1r^2 + q_2r^4 + q_3r^6\right) + \left[p_1\left(r^2 + 2\bar{x}^2\right) + 2p_2\bar{x}\bar{y}\right]\left(1 + p_3r^2\right) \\ \mathrm{d}y = \bar{y}\left(q_1r^2 + q_2r^4 + q_3r^6\right) + \left[p_2\left(r^2 + 2\bar{y}^2\right) + 2p_1\bar{x}\bar{y}\right]\left(1 + p_3r^2\right) \end{cases} \tag{2.19}$$

Here, $\bar{x}$, $\bar{y}$ and r can be defined as follows:

$$\begin{cases} \bar{x} = X - u_0 \\ \bar{y} = Y - v_0 \\ r^2 = \bar{x}^2 + \bar{y}^2 \end{cases} \tag{2.20}$$

To sum up, in the imaging model of the star sensor, linear model parameters f, (u_0, v_0) and nonlinear distortion coefficient $(q_1, q_2, q_3, p_1, p_2, p_3)$ constitute the intrinsic parameters of star sensor, while $(r_1$–$r_9)$ makes up the extrinsic parameter.

In order to conduct a subsequent performance evaluation of the identification rate of the star identification algorithms, a certain level of positional noise (a gauss noise with mean = 0 and variances σ_x, σ_y) is added to the coordinates of the star spot image on the focal plane of the star sensor in a star image simulation so as to simulate centroid error.

2.3.2 The Composition of the Digital Star Image

For star sensor, the star can be viewed as a point light source. The positioning accuracy of a single pixel cannot meet the demand for attitude establishment. Thus, defocusing is often used to make the star spot image spread to multiple pixels, and then centroiding methods are used to obtain sub-pixel positioning accuracy [1]. The pixel size of the star spot is not only related to a star's brightness (magnitude), but also related to the PSF of the optical system. The gray distribution of a star spot image follows the PSF of the optical system and can be approximately represented by a two-dimensional Gaussian distribution function:

$$\mu_i(x, y) = \frac{1}{\sqrt{2\pi}\sigma}\exp\left(-\frac{(x - \bar{X}_i)^2 + (y - \bar{Y}_i)^2}{2\sigma^2}\right) \tag{2.21}$$

Assume that there are N stars in the star image, and a photoelectron density that is formed in the imaging process of target star spot can be expressed as follows [9]:

$$s(x, y) = \sum_{i=1}^{N} k \int s_i A\tau(\lambda) Q\mu_i(x, y) P(\lambda) t_s \mathrm{d}\lambda \tag{2.22}$$

Here, k stands for bandwidth impact factor, A for optical entrance pupil area, $\tau(\lambda)$ for optical transmittance, and Q for quantum efficiency. $P(\lambda)$ stands for the spectral response of the imaging device and t_s for integration time.

$$s_i = 5 \times 10^{10}/2.512^{M_i}.$$

In the above equation, M_i stands for the magnitude of the i-th star.

The photoelectron density that is formed in the process of background imaging can be expressed as follows:

$$b = \int b_0 A\tau(\lambda)P(\lambda)A_p t_s \mathrm{d}\lambda \tag{2.23}$$

Here, $b_0 = 5 \times 10^{10}/2.512^{M_b}$, M_b stands for the magnitude order of the background which can be generally of brightness of 10.0 Mv. A_p stands for angle area of a single pixel.

Thus, the total number of photoelectrons acquired by the (m, n)-th pixel $(0 \le m < M,\ 0 \le n < N)$ on the photosensitive surface is as follows:

$$I(m, n) = \int_{m\mathrm{d}x}^{(m+1)\mathrm{d}x} \int_{n\mathrm{d}y}^{(n+1)\mathrm{d}y} (s(x, y) + b)\mathrm{d}x\mathrm{d}y \tag{2.24}$$

Integrate Eqs. (2.22) and (2.23), and then the above equation can be simplified as:

$$I(m, n) = \int_{m\mathrm{d}x}^{(m+1)\mathrm{d}x} \int_{n\mathrm{d}y}^{(n+1)\mathrm{d}y} \sum_{i=1}^{N} \frac{C}{2.512^{M_i}} \exp\left(-\frac{(x - \bar{X}_i)^2 + (y - \bar{Y}_i)^2}{2\sigma^2} \right) \mathrm{d}x\mathrm{d}y + B \tag{2.25}$$

Here, both B and C are constants.

In addition to the background and star spot, the final star image should also include noise signal which mainly consists of shot noise and the dark current noise of the image device. The white noise model can be represented by the random number of Gaussian distribution:

$$N(m, n) = \mathrm{normrand}(0, \sigma_N) \tag{2.26}$$

Thus, the final output of the image signal can be expressed as follows:

$$P(m, n) = I(m, n) + N(m, n) \tag{2.27}$$

Table 2.3 Simulation parameters of star image

	Parameter values
Pixel resolution	1024 × 1024
Pixel size	12 μm × 12 μm
Position of the principal point	(512,512)
Focal length of optical system	58.4536 mm
FOV	12° × 12°
Maximum magnitude sensitized	6 Mv
Radius of PSF	1 pixel

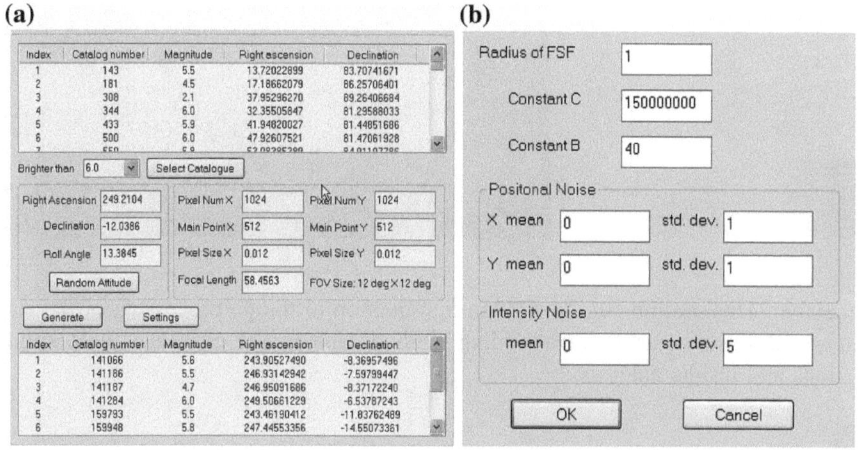

Fig. 2.10 Parameter setting in star image simulation

The parameters used in the process of star image simulation are shown in Table 2.3. In the process of image synthesis, parameters B, C, σ_x, σ_y, and σ_N can take the appropriate values as required based on the design specifications of the optical system and image device as shown in Fig. 2.10a, b. Figure 2.11 is a star image simulated when the attitude angle of star sensor is (249.2104, −12.0386, 13.3845).

2.4 Star Spot Centroiding

Restricted by the manufacturing technology of the imaging device, image resolution cannot be improved indefinitely. Thus, the method of increasing positioning accuracy by improving image resolution is limited. However, the method of image processing carried out by using software to conduct star spot centroiding is

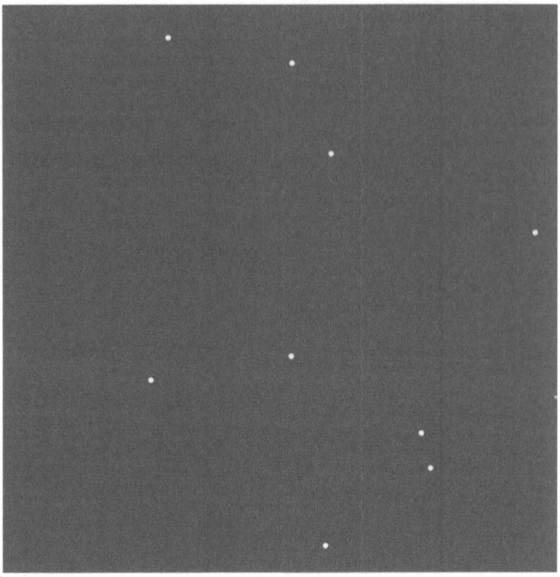

Fig. 2.11 Simulated star images

effective. This section begins with an introduction to the preprocessing of the star image, discusses star spot centroiding methods and finally concludes with simulations and results analysis.

2.4.1 Preprocessing of the Star Image

A salient feature of low-level processing of a star image is the large amount of data. Take the eight-bit gray image with 1024×1024 resolution for example. The amount of data in each frame is 1 MB. Thus, in order to achieve real-time processing, the low-level processing algorithms of the star image must not be too complex. Meanwhile, considering the requirement that the low-level processing of a star image be achieved by adopting specific hardware circuit (e.g., FPGA or ASIC), the algorithms must also be characterized by parallel processing as much as possible. Preprocessing of the star image mainly includes noise removal processing and rough judgment of the stars.

The output of the image sensor is done via an image capture circuit. The original digital image signal obtained in this way is mixed with a lot of noises. Thus, noise removal processing of the original star image is generally done first. The common noise removal processing can use 3×3 or 5×5 low-pass filter template, for example, neighborhood average template and a Gauss template. To reduce the amount of computation of the algorithms, 3×3 low-pass filter template is often used.

Before extracting the information of star spot positions, the target star spots in the star image must be roughly judged. Rough judgment is actually a process of image segmentation which can be divided into two stages:

① separating the target star spot from the background;
② separating a single target star spot from others.

At the first stage, global threshold or local threshold can be used to segment the image. Generally, considering the characteristics of the star image and the complexity of algorithms, just one fixed global background threshold can be used to separate the star spot from the background. The selection of the global threshold can adopt multi-window sampling, i.e., selecting several windows randomly in the image, computing the mean of their gray distribution, and then taking this value as the mean of the background's gray distribution. Generally, the background mean plus five times the standard deviation of noise can be treated as the global background threshold [1].

How to separate one target star spot from another, is a problem in the preprocessing of a star image. Ju [3] uses multi-threshold clustering to conduct clustering identification of the pixels whose gray value is greater than the global background threshold. The specific procedures are as follows:

① Set ten thresholds, group the pixels whose gray value is greater than the global background threshold together into the corresponding interval based on their gray values, and then put them in order.
② Each interval is scanned in descending order, the pixel with the maximum gray value is found in the current interval, and its neighborhood spot is sought out. They are regarded as belonging with the same star.

This method is relatively complex and involves a sorting operation. Considering the specific requirement for speed in star image processing, binary image processing can be used for reference, and then the connected domain algorithms [10] can be used to achieve a clustering judgment of the star spot. And the specific procedures are as follows:

① The image is scanned from left to right, top to bottom.
② If the gray value of a pixel is greater than background threshold (T), then:

  * If there is a marker in the above or left spots, copy this marker.
  * If there is the same marker in the above or left spots, copy this marker.
  * If there are different markers in the above or left spots, copy the marker of the above spots and then the two markers are put into the equivalent table as an equivalent marker.
  * Otherwise, allocate a new marker to this pixel and put it into the equivalent table.

③ Repeat step ② until all the pixels whose gray value is greater than T are scanned.

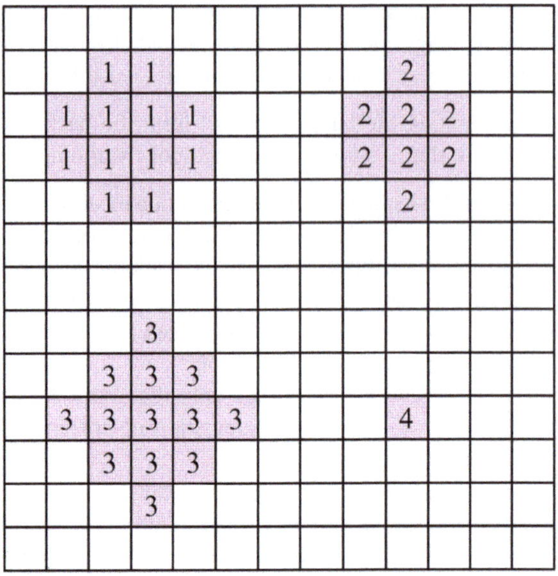

Fig. 2.12 Connected domain segmentation algorithms

④ Combine the pixels with the same marker in the equivalent table and reallocate a marker with a lower index number to them.

After the segmentation based on connected domains algorithm, each star spot is represented by a set of neighboring pixels with the same marker, for example, spots 1, 2, 3 in Fig. 2.12. To eliminate the influence of potential noise interference, star spots whose number of pixels is lower than a certain threshold should be abandoned, for example, spot 4 in Fig. 2.12. As can be seen from the above procedures, threshold segmentation and connected domain segmentation algorithms can be done at the same time. Thus, the image can be scanned just once, which is very suitable for realization by a specific hardware circuit and can meet real-time demands [11].

2.4.2 Centroiding Methods

Generally, many current centroiding methods can achieve sub-pixel (or even higher) accuracy. To obtain higher accuracy in star spot positions from the star image, defocusing is often used to make the imaging points of stars on the photosensitive surface of image sensor spread to multiple pixels. Both theoretical derivations and experiments prove that ideal centroiding accuracy can be achieved when the diameter of dispersion circle ranges from three to five pixels [12, 13].

There are two categories of centroiding methods for spot-like image: the one based on gray and the one based on edge [14]. The former often uses the information of spot's gray distribution, for example, centroid method, surface fitting method, etc. The latter often uses the information of a spot's edge shape, for example, edge circle (ellipse) fitting, Hough transformation, etc. The former applies to relatively small spots with an even distribution of gray, while the latter applies to larger spots that are less sensitive to gray distribution. Generally, the diameter of spot star spots in the actual measured star image ranges from three to five pixels and their gray values approximately follow a Gaussian distribution. Thus, for target star spots, it is more appropriate to adopt the methods based on gray to conduct centroiding processing. Simulation experiments also show that the accuracy of this method is higher than that of the method based on the edge method. Here, the former is mainly introduced, including the centroid method, the modified centroid method and the Gaussian surface fitting method. Then their positioning accuracy is analyzed.

(1) Centroid Method

Assume the image that contains target star spots is represented by $f(x, y)$. Here,

$$x = 1, \ldots, m, \quad y = 1, \ldots, n.$$

The process of thresholding is as follows:

$$F(x, y) = \begin{cases} f(x, y) & f(x, y) \geq T \\ 0 & f(x.y) < T \end{cases} \tag{2.28}$$

In the above equation, T stands for background threshold. Centroid method is actually the first moment of the image after thresholding:

$$x_0 = \frac{\sum_{x=1}^{m} \sum_{y=1}^{n} F(x, y)x}{\sum_{x=1}^{m} \sum_{y=1}^{n} F(x, y)}, \quad y_0 = \frac{\sum_{x=1}^{m} \sum_{y=1}^{n} F(x, y)y}{\sum_{x=1}^{m} \sum_{y=1}^{n} F(x, y)} \tag{2.29}$$

The centroid method is the most commonly used. It is easy to realize with a relatively high positioning accuracy. It requires that the gray distribution of the spot image be relatively even. It has some modified forms, including a centroid method including a threshold and square weighting centroid method.

(2) Square Weighting Centroid Method

The computational equation of the square weighting centroid method can be expressed as follows:

$$x_0 = \frac{\sum_{x=1}^{m} \sum_{y=1}^{n} F^2(x, y)x}{\sum_{x=1}^{m} \sum_{y=1}^{n} F^2(x, y)}, \quad y_0 = \frac{\sum_{x=1}^{m} \sum_{y=1}^{n} F^2(x, y)y}{\sum_{x=1}^{m} \sum_{y=1}^{n} F^2(x, y)} \tag{2.30}$$

The square weighting centroid method substitutes the square of the gray value for the gray value expressed as weight. It highlights the influence of the pixel which is closer to the center and with a relatively large gray value on the central position.

(3) Centroid Method with Threshold

$F(x,y)$ in Eq. (2.28) is redefined as follows by centroid method with threshold [15, 16]:

$$F'(x,y) = \begin{cases} f(x,y) - T & f(x,y) \geq T' \\ 0 & f(x,y) < T' \end{cases} \tag{2.31}$$

Here, T' is the selected threshold. Generally, $T' > T$. The computational equation of modified centroid method is as follows:

$$x_0 = \frac{\sum_{x=1}^{m}\sum_{y=1}^{n} F'(x,y)x}{\sum_{x=1}^{m}\sum_{y=1}^{n} F'(x,y)}, \quad y_0 = \frac{\sum_{x=1}^{m}\sum_{y=1}^{n} F'(x,y)y}{\sum_{x=1}^{m}\sum_{y=1}^{n} F'(x,y)} \tag{2.32}$$

This method is to find the centroid of pixels with thresholds greater than T' in the original image, equivalent to the original image minus the background threshold. It can be proved that the centroid method with threshold is of a higher accuracy than the traditional centroid method. When $T' = T$ and the gray distribution $f(x,y)$ is not related to the coordinate values of x and y is the centroid method with threshold equivalent to the traditional centroid method.

(4) Surface Fitting Method

Since images of stars on the photosensitive surface of the image sensor can be approximately viewed as of a Gaussian distribution, a Gaussian surface can be used to fit the gray distribution. The two-dimensional Gaussian surface function can be expressed as follows:

$$f(x,y) = A \cdot \exp\left\{-\frac{1}{2(1-\rho^2)}\left[\left(\frac{x-x_0}{\sigma_x}\right)^2 - 2\rho\left(\frac{x}{\sigma_x}\right)\left(\frac{y}{\sigma_y}\right) + \left(\frac{y-y_0}{\sigma_y}\right)^2\right]\right\} \tag{2.33}$$

Here, as scale coefficient, A stands for the size of gray amplitude. It is related to the brightness (magnitude) of stars. (x_0, y_0) stands for the center of Gaussian function, σ_x, σ_y for the standard deviation in the directions of x and y, respectively, and ρ for the correlation coefficient. Generally, take $\rho = 0$ and $\sigma_x = \sigma_y$. The center (central position coordinates of stars) of the Gaussian function can be obtained by the least square method. To facilitate computation, one-dimensional Gaussian curves in the directions of x and y can be used for fitting, respectively.

The one-dimensional Gaussian curve equation is as follows:

$$f(x) = A \cdot \mathrm{e}^{-\frac{(x-x_0)^2}{2\sigma^2}} \tag{2.34}$$

Take the logarithm of Eq. (2.34) and the result is as follows:

$$\ln f(x) = a_0 + a_1 x + a_2 x^2 \tag{2.35}$$

Here,

$$a_0 = \ln A - x_0^2/2\sigma^2,\ a_1 = x_0/\sigma^2,\ a_2 = -1/2\sigma^2$$

Fit the points in the direction of x, and coefficients a_0, a_1, a_2 of the above quadratic polynomials can be acquired by the least square:

$$x_0 = -\frac{a_1}{2a_2} \tag{2.36}$$

Equation (2.36) is the x coordinate of the central point.

Similarly, the y coordinate of the central point can be obtained.

2.4.3 Simulations and Results Analysis [17]

To verify the accuracy of various centroiding methods, star spot images can be generated based on digital image simulations introduced in Sect. 2.3. The image size is 20 × 20. There is only one star in the image and the central position of the star spot is (10, 10). The radius of PSF can take one pixel, and the background's gray value of the image is 20. To investigate the influences of gray noise and spot image size on positioning accuracy, the standard deviation of gray noise varies from zero to ten and the magnitude from one to six. Simulation experiments use stars of 5.5 Mv as references. The maximum gray value of its peak point just reaches saturation, i.e., 255. Figure 2.13a shows a star image. Its standard deviation of noise is eight, and its radius of PSF is 1.5. Figure 2.13b shows the amplified image of the original star spot.

(1) Influence of Gray Noise on Positioning Accuracy

Assume the actual central coordinates of the star spot are (x_c, y_c) and the measured central coordinates are (x_i, y_i). The deviation (e_p) of centroiding and the standard deviation (σ_p) are defined as follows:

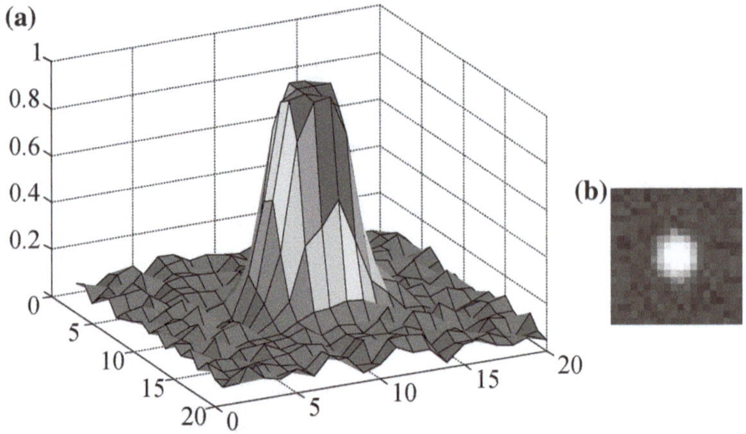

Fig. 2.13 Star images simulated in simulation experiments

$$
\begin{aligned}
e_p &= \frac{1}{n}\sum_{i=1}^{n}\sqrt{(x_i - x_c)^2 + (y_i - y_c)^2} \\
\sigma_p &= \sqrt{\frac{1}{n-1}\sum_{i=1}^{n}\left(\sqrt{(x_i - x_c)^2 + (y_i - y_c)^2} - e_p\right)^2}
\end{aligned}
\tag{2.37}
$$

Here, $i = 1, \ldots, n$, with n for the times of measurements.

The accuracy of various centroiding methods changes with the variance of Gauss noise. The positioning accuracy changes of several centroiding methods are illustrated in Fig. 2.14. It shows the comparison curves when the standard deviation of gauss noise varies from zero to ten. In the simulation experiment, the radius of PSF

Fig. 2.14 Influence of gray noise on positioning accuracy

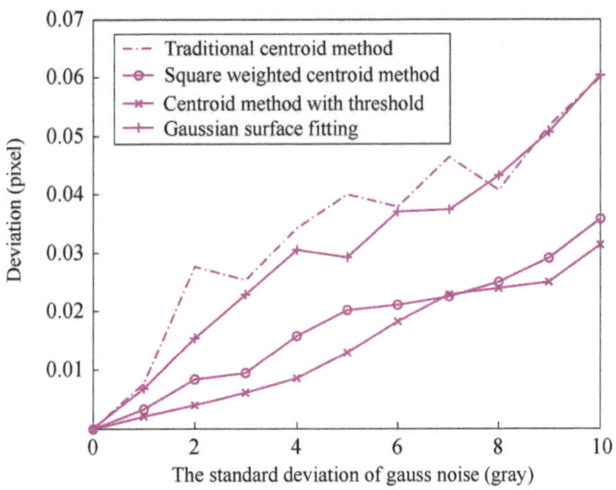

takes one pixel. Binary threshold T is equal to the background's gray value plus five times the standard deviation of gauss noise. The selected threshold T' of centroid method with threshold is equal to T plus 20 ($T' = T + 20$). Each method undergoes 1000 times of measurements.

The following conclusions can be drawn from the above:

① When noise level is low, each method is of high accuracy and the accuracy is nearly the same.
② Increase the noise level and the accuracy of various methods decreases. The accuracy decrease of the traditional centroid method is most significant, while the decrease of centroid method with threshold is relatively small.
③ The accuracy comparison of various methods is as follows:
centroid method with threshold > square weighting centroid method > Gaussian surface fitting method > traditional centroid method.

(2) Influence of Noise Removal Processing on Positioning Accuracy

As shown in Fig. 2.14, the accuracy of various centroiding methods decreases to varying degrees when the noise level is high. Thus, it is feasible to conduct a low-pass filter processing and then centroiding of star images with noise. And a 3 × 3 neighborhood average template can be used to conduct low-pass filter processing of images so as to investigate the influence of noise removal on positioning accuracy. After noise removal processing, the deviation comparison curves of various centroiding methods are shown in Fig. 2.15.

By comparing Figs. 2.14 and 2.15, it can be seen that, after a low-pass filter processing, the positioning accuracy of various methods is improved to some degree. Low-pass filter processing can significantly improve the accuracy of traditional centroid methods, but has little impact on other methods. Thus, when other methods are used to conduct centroiding, filter and noise removal in preprocessing can be omitted to save time.

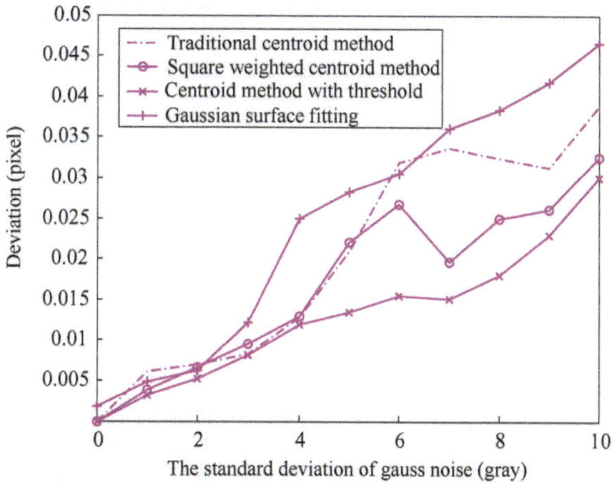

Fig. 2.15 Influence of gray noise on positioning accuracy after noise removal

(3) Influence of Magnitude on Positioning Accuracy

Magnitude determines the size of star spot image. The star spot images (amplified) formed by stars of 6–1 Mv are shown in Fig. 2.16a. Figure 2.16b shows the comparison curves of positioning accuracy obtained by conducting centroiding of the images, respectively. In the experiment, the standard deviation of gray noise is 5.

As shown in Fig. 2.16b, with the increase of magnitude, the positioning accuracy of the centroid method and the modified centroid method is improved, though by not very much. When the Gaussian surface fitting method is used, the deviation

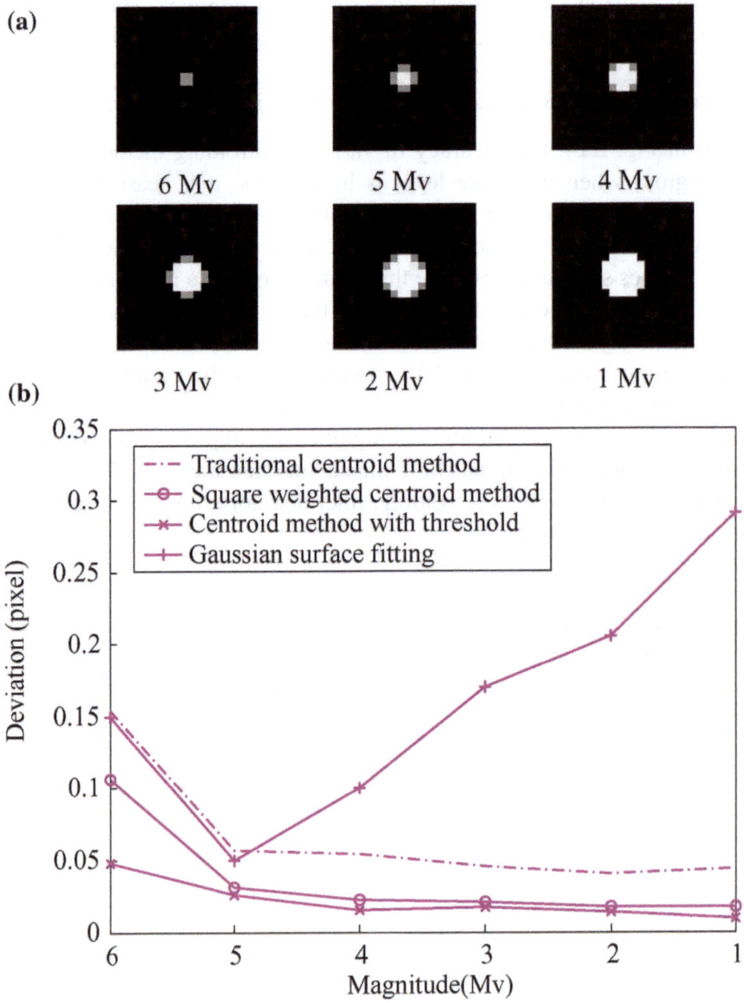

Fig. 2.16 Influence of magnitude on positioning accuracy

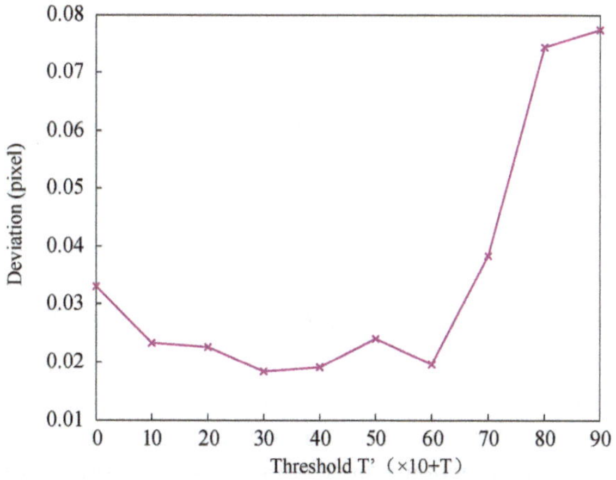

Fig. 2.17 Influence of threshold T' on positioning accuracy

first decreases and then increases. The main reason for this is that the gray distribution of star spots reaches saturation and can no longer be fitted through the Gaussian surface. From the above simulations and results analysis, it can be concluded that the centroid method with threshold is a centroiding method that is appropriate for extracting the central positions of star spots. It is of high accuracy and is robust to the influence of noise. In addition, the centroid method with threshold is as simple as the traditional centroid method and is easily implemented.

(4) Selection of Threshold

Threshold T' is an important parameter for the centroid method with threshold, the selection of which affects positioning accuracy. Figure 2.17 shows the variation curve of positioning accuracy with different values of T'. As shown in Fig. 2.17, the positioning accuracy is nearly the same when T' is between $T + 10$ and $T + 60$. Within this range, the deviation reaches a minimum value when T' is approximately $T + 30$.

When the standard deviation of gray noise is 8, $T' = T + 20$, Magnitude = 5, and there is no filter processing, the deviation of the centroid method with threshold is 0.023 pixel and the standard deviation is 0.012 pixel. Thus, it can be concluded that this centroiding method can achieve the positioning accuracy of 0.04–0.05 pixel.

2.5 Calibration of Centroiding Error

The field angle of star is far less than one arc second. Ideally, focus imaging leaves the star spot image of the star sensor within one pixel. To improve the centroiding accuracy of the star sensor, a defocusing technique is often used to make the star spot image spread to a dispersion circular spot. And then the centroid algorithms are used to compute the center of star spot so as to obtain sub-pixel centroiding accuracy. Currently, the centroiding accuracy of the star sensor can generally achieve 1/10–2/20 pixel. To achieve centroiding of higher accuracy, optimization design and noise suppression have to be done for an imaging driving circuit. In addition, based on the characteristics of star imaging and the working characteristics of image device, centroiding error has to be compensated more precisely.

2.5.1 Pixel Frequency Error of Star Spot Centroiding

There are many factors affecting the accuracy of the centroid algorithms, including noise, sampling and quantization errors, etc. These factors, based on their influence form and function, can be put into two categories: those that cannot be compensated and those that can be compensated. Generally speaking, it is very difficult to compensate all kinds of noises (e.g., readout noise, dark current noise, fixed pattern noise, nonuniformity noise, etc.) later in the imaging process. Instead, specific measures are often taken when the image capture circuit is designed to improve the signal-to-noise ratio (SNR) as much as possible. Factors such as sampling and quantization errors have a regular influence on centroiding. In itself, it is an error introduced when an energy distribution center is replaced by a pixel geometric center. This kind of error is often called pixel frequency error. That is, the deviation of the star spot center changes regularly within one pixel [18–20].

(1) Centroiding Deviation Induced by Fill Factor

Generally, pixel fill factor is assumed to be 100% in the centroiding process. In fact, since the pixel fill factor of the image device is less than 100%, nonuniformity of pixel response in space is caused in the quantization process of pixel. Even if there is no noise, pixel quantization will still result in the distortion of PSF and the deviation between the calculated star spot position and its "real" position.

A star spot covering 3×3 pixels (as shown in the circular region of the dotted line) which moves in the direction of the line scanning is illustrated by Fig. 2.18. The dark rectangles in Fig. 2.18 are the regions occupied by transistors of reset, row selection gating, and column amplification. And the surrounding white regions are the effective photosensitive parts of pixels. In the process of scanning, with the changes in pixel photosensitive regions (shaded parts in the circular region of dotted line in Fig. 2.18) covered by the spot, there appears to be a periodical change in the computed spot centroid deviation.

(2) Centroiding Deviation Induced by Charge Saturation

Sampling and quantization errors induced by charge saturation are another important source. When saturation of electron charge occurs in the center pixel of the center energy of star spot, it causes the traditional Gaussian PSF model to be truncated. And the truncation effect induces computational deviation in star spot

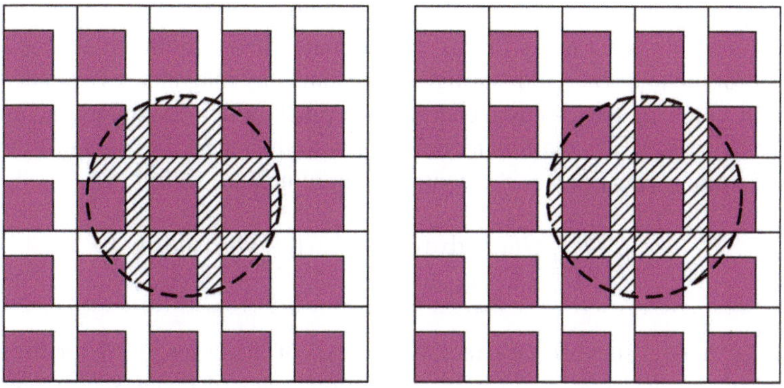

Fig. 2.18 Influence of fill factor on centroiding accuracy

centroid. The truncation of PSF in the direction of X is illustrated in Fig. 2.19. With pixels whose gray value is greater than 255, PSF is then truncated. The truncated Gaussian PSF model is as follows:

$$I(x,y) = \begin{cases} I_0 \exp\left(-\left((x-x_0)^2 + (y-y_0)^2\right)/(2\sigma^2)\right) \Big/ (2\pi\sigma^2) & x^2+y^2 \geq r^2 \\ I_1 & x^2+y^2 < r^2 \end{cases} \tag{2.38}$$

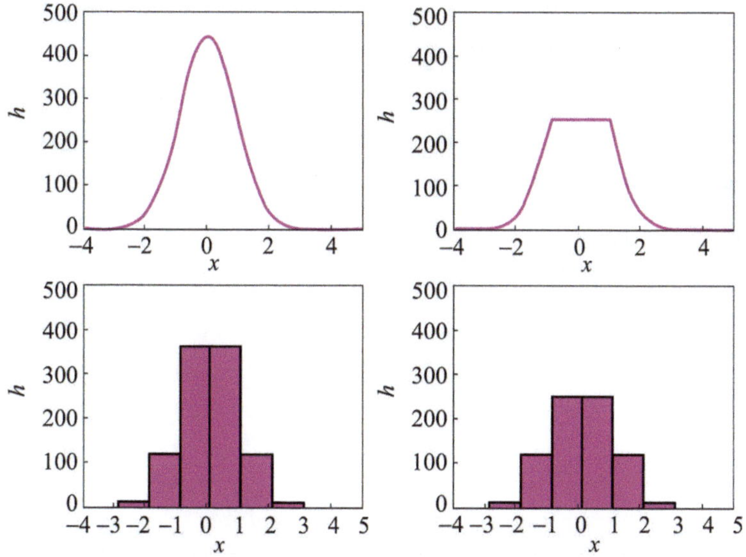

Fig. 2.19 Illustration of Gaussian PSF truncation model induced by charge saturation

Here, I stands for the radiation energy distribution of star spot, x_0 and y_0 for the center of the star spot, x and y for pixel coordinates, and r for the truncation radius. I_0 stands for the total energy of starlight, determined by magnitude. I_1 stands for the saturation value of the electron charge in a pixel. When magnitude is low and starlight is weak, $r = 0$ and this model returns to the original Gaussian PSF.

The common equations of the centroid algorithms are as follows:

$$x_c = \frac{\int\int_A xI(x,y)\mathrm{d}x\mathrm{d}y}{\int\int_A I(x,y)\mathrm{d}x\mathrm{d}y}, \quad y_c = \frac{\int\int_A yI(x,y)\mathrm{d}x\mathrm{d}y}{\int\int_A I(x,y)\mathrm{d}x\mathrm{d}y} \tag{2.39}$$

Here, x_c and y_c stand for the radiation center of the star spot, A for the neighborhood of the star spot, x and y for pixel coordinates of image sensor plate, and $I(x, y)$ for radiation distribution function. After discretization of digital images, the computational equations of centroid are as follows:

$$\tilde{x}_c = \frac{\sum_{k=1}^{n} x_k I_k}{\sum_{k=1}^{n} I_k}, \quad \tilde{y}_c = \frac{\sum_{k=1}^{n} y_k I_k}{\sum_{k=1}^{n} I_k} \tag{2.40}$$

Here, $\tilde{x}_c$ and $\tilde{y}_c$ stand for the center of star spot after discretization, n for the number of pixels with a gray value greater than threshold T', k for the index number of pixels, x_k and y_k for the coordinates of the k-th pixel, and I_k for the gray output of the k-th pixel.

Figure 2.20 shows the pixel frequency error curves induced by charge saturation obtained through simulations with different magnitudes when there is no noise. The directions of x and y are mutually independent, and here the error in the direction of x is simulated. As shown in Fig. 2.20, the deviation within pixels approximately follows the sine function. But with the changes of magnitude, from 3.5 to 0.0 Mv, both the amplitude and phase change.

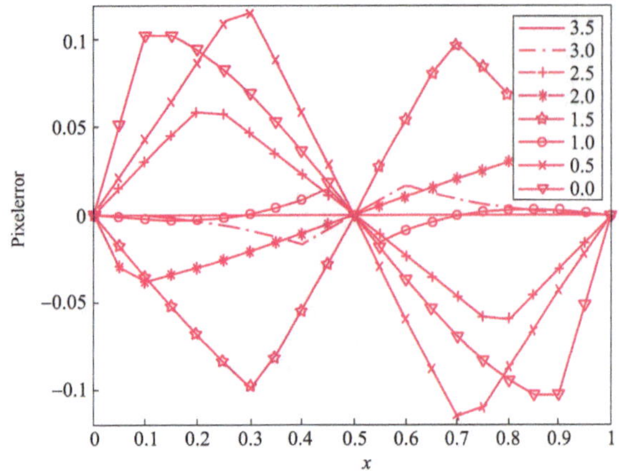

Fig. 2.20 Centroiding pixel frequency error of star spots with different magnitudes induced by charge saturation

2.5.2 Modeling of Pixel Frequency Error

Next, with the star of 0 Mv as reference, the pixel frequency error model of the centroid algorithms is introduced, and brightness and positional noise are increased to simulate and compute pixel frequency error. First, the pixel frequency error of the model built is as follows:

$$\begin{cases} E_x = A_x\left(\sin\left(2\pi x_p + 2\pi B_x\right) - \sin(2\pi B_x)\right) \\ E_y = A_y\left(\sin\left(2\pi y_p + 2\pi B_y\right) - \sin\left(2\pi B_y\right)\right) \end{cases} \tag{2.41}$$

Here, A_x and A_y stand for the deviation amplitude coefficients in the directions of x and y, B_x and B_y for deviation phase coefficients, and x_p and y_p for position coordinates within one pixel. Since the pixel of the image sensor is square shaped, assume $A_x = A_y$ and $B_x = B_{y.}$ And for the same magnitude, the deviation amplitude coefficient and phase coefficient are constants. Here, use the direction of x as reference, and the parameter estimation in the direction of y can be processed in the same way.

Using least square estimation, the estimation equations of amplitude and phase deviations are as follows:

$$\begin{cases} \Delta E_x = \left[\sin\left(2\pi x_p + 2\pi B_x\right),\quad 2\pi\cos\left(2\pi x_p + 2\pi B_x\right) - 2\pi\cos(2\pi B_x)\right]\begin{bmatrix}\Delta A_x \\ \Delta B_x\end{bmatrix} \\ \Delta E_y = \left[\sin\left(2\pi y_p + 2\pi B_y\right),\quad 2\pi\cos\left(2\pi y_p + 2\pi B_y\right) - 2\pi\cos\left(2\pi B_y\right)\right]\begin{bmatrix}\Delta A_y \\ \Delta B_y\end{bmatrix} \end{cases} \tag{2.42}$$

Here, ΔE_x and ΔE_y stand for the measured values of pixel deviation in the directions of x and y, ΔA_x and ΔB_x for pixel frequency error amplitude and phase estimation in the direction of x, and ΔA_y and ΔB_y for pixel frequency error amplitude and phase estimation in the direction of y.

2.5.3 Calibration of Pixel Frequency Error

The basic parameters of the simulation star sensor are as follows:

FOV: 12° × 12°
pixel array: 1024 × 1024
pixel size: 0.015 mm × 0.015 mm
focal length: 73.03 mm.

Assume the random error of the central position of the star spot within pixels is Gaussian white noise with mean = 0 and mean square deviation = 0.01 pixel. The

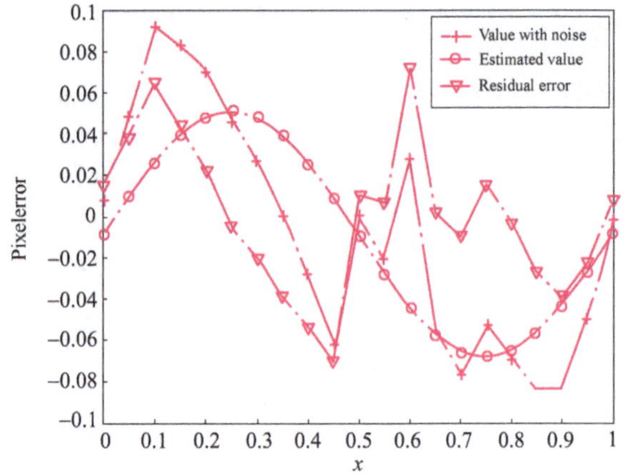

Fig. 2.21 Calibration of pixel frequency error

brightness error is also Gaussian white noise and its mean square deviation is 2% of the saturated gray value. The threshold takes five times the mean square deviation of brightness, i.e., 10% of the saturated gray value. Figure 2.21 shows the calibration results of pixel frequency error. '−+' stands for the simulation value with noise, '−O' for sine estimated value, and '−Δ' for residual error.

The amplitude of sine deviation after calibration is 0.060 pixel and the phase is $1.4 \times 10^{-3\ \mathrm{rad}}$. The mean root of the simulation value's error is 0.055 pixel, and the mean root of residual error after calibration is 0.036 pixel. The computation accuracy of star spot centroid is improved by 34%, which shows the calibration value is of remarkable accuracy. It is worth noting that the above simulations are based on the star of 0 Mv. The pixel frequency error of each magnitude is different. Thus, to fully calibrate the pixel frequency errors of all magnitudes to be dealt with by the star sensor, the amplitude and phase of several major magnitudes' pixel frequency error can be measured, and then interpolation can be used to calibrate the centroiding results of different magnitudes.

In the laboratory, the starlight simulator and a turntable of high accuracy can be used to calibrate the pixel frequency error of centroiding. As shown in Fig. 2.22, the turntable is adjusted to leave the central positions of the imaging star spot on the edge of pixels to reduce possible interferences in the direction of y. The turntable begins from the initial position of pixel x. Its rotation angle from pixel x to pixel $x + 1$ is approximately 0.05 pixel each interval. In the process, 21 points are

Fig. 2.22 Sampling within a pixel

sampled. Conduct multiple samplings (100 times) at each point to reduce random error. Repeat the sampling process in the direction of y. To obtain a more precise estimated parameter value of pixel frequency error, five pixels can be selected in the up, down, left, right, and middle parts of the star sensor's image sensor plate to repeat the above data collection. The parameters can be integrated to solve the amplitude and phase deviation coefficients. For example, for captured image of resolution 1024 × 1024, the sampling of pixel frequency error can be done at five pixel points, i.e., (127, 512), (512, 512), (896, 512), (512, 127), (512, 896), respectively.

References

1. Liebe CC (2002) Accuracy performance of star trackers - a tutorial. IEEE Trans Aerosp Electron Syst 38(2):587–599
2. Jeffery WB (1994) On-orbit star processing using multi-star star trackers. SPIE 2221:6–14
3. Ju G, Kim H, Pollock T et al (1999) DIGISTAR: a low-cost micro star tracker. AIAA Space Technology Conference and Exposition, Albuquerque, AIAA: 99-4603
4. Chen Y (2001) A research on three-axis attitude measurement of satellites based on star sensor. Doctoral Thesis of Changchun Institute of Optics, Fine Mechanics and Physics, Chinese Academy of Sciences, Changchun
5. Wei X (2004) A research on star identification methods and relevant technologies for star sensor (pp 15–43). Doctoral Thesis of Beijing University Aeronautics and Astronautics, Beijing
6. Zhang G, Wei X, Jiang J (2006) Star map identification based on a modified triangle algorithm. Acta Aeronautica et Astronautica Sinica 27(6):1150–1154
7. Wang X (2003) Study on wild-field-of-view and high-accuracy star sensor technologies. Doctoral Thesis of Changchun Institute of Optics, Fine Mechanics and Physics, Chinese Academy of Sciences, Changchun
8. Weng JY (1992) Camera calibration with distortion models and accuracy evaluation. IEEE Trans Pattern Anal Mach Intell 14(10):965–980
9. Yuan J (1999) Navigation star sensor technologies. Doctoral Thesis of Sichuan University, Chengdu
10. Zhang Y (2001) Image segmentation. Science Press, Beijing
11. Hao X, Jiang J, Zhang G (2005) CMOS star sensor image acquisition and real-time star centroiding algorithm. J Beijing Univ Aeronaut Astronaut 31(4):381–384
12. Grossman SB, Emmons RB (1984) Performance analysis and optimization for point tracking algorithm applications. Opt Eng 23(2):167–176
13. Zhou R, Fang J, Zhu S (2000) Spot size optimization and performance analysis in image measurement. Chin J Sci Instrum 21(2):177–179
14. Shortis MR, Clarke TA, Short TA (1994) Comparison of some techniques for the subpixel location of discrete target images. SPIE 2350:239–250
15. Sirkis J (1990) System response to automated grid methods. Opt Eng 29(12):1485–1491
16. West GAW, Clarke TA (1990) A survey and examination of subpixel measurement techniques. SPIE 1395:456–463
17. Wei X, Zhang G, Jiang J (2003) Subdivided locating method of star image for star sensor. J Beijing Univ Aeronaut Astronaut 29(9):812–815
18. Giancarlo R, Domenico A (2003) Enhancement of the centroiding algorithm for star tracker measure refinement. Acta Astronaut 53:135–147

19. Ying JJ, He YQ, Zhou ZL (2009) Analysis on laser spot locating accuracy affected by CMOS sensor fill factor in laser warning system. The ninth international conference on electronic measurement and instruments (pp 202–206) (2)
20. Hao X (2006) CHU key technologies of miniature CMOS star sensor (pp 61–64). Doctoral Thesis of Beijing University Aeronautics and Astronautics, Beijing

Chapter 3
Star Identification Utilizing Modified Triangle Algorithms

Generally, the existing star identification algorithms can be roughly divided into two classes according to their methods of feature extraction: subgraph isomorphism algorithms and star pattern recognition algorithms. The former category regards star pair angular distances as sides and stars as vertexes, so that a measured star image can be regarded as the subgraph of a full-sky star image. By using angular distance in a direct or indirect manner, these algorithms use line (angular distance), triangle, and quadrangle as the basic matching elements to build a guide database in a certain way. With the combined use of those elements, once the only area (subgraph) that meets the matching requirements is found in the full-sky star image, it is regarded as the corresponding match of the measured star image. The triangle algorithm is the most typical subgraph isomorphism algorithm and has been so far one of the most common and widely used star identification algorithms. For example, the space-borne star sensors of the Danish Oersted satellite [1, 2], America's DIGISTAR I miniature star tracker [3], and others all use this algorithm. Besides, the triangle algorithm has many derived forms, like Scholl [4] method based on six features and Mortari et al.'s [5] pyramid algorithm.

Traditional triangle algorithms, though simple in structure and easy in realization, have many weaknesses. For example, they generally require very large guide databases and are rather time-consuming in data retrieval. The modification of the triangle algorithms mainly focuses on the introduction of new features, especially that on magnitude, and the selection of triangles to eliminate as many redundant guide triangles as possible. Due to the algorithms' own limitations, the modifications cannot remarkably improve the performance. Zhang et al. [6–9] modify the current triangle algorithms for star identification so as to solve their existing problems and enhance the efficiency of matching and identification.

In this chapter, two modified triangle algorithms for star identification are introduced, and their basic principles and the specific realization processes are elaborated on. Finally, the performance of the two algorithms will be evaluated through experiments and compared with that of the traditional triangle algorithm.

G. Zhang, *Star Identification*, DOI 10.1007/978-3-662-53783-1_3

3.1 Current Triangle Algorithms

The triangle algorithms for star identification, with many varied forms, use angular distances between each two of the three stars to fulfill matching and identification. In this section, their basic principles are introduced, and their features and existing problems analyzed.

3.1.1 Basic Principles of the Triangle Algorithm

A single star cannot be used for identification, while two stars can be identified through angular distance. The right ascension and declination coordinates of guide stars *i* and *j* are denoted as (α_i, δ_i) and (α_j, δ_j), respectively. The angular distance in the celestial coordinate system is defined as follows (c.f. Fig. 3.1a):

$$d(i,j) = \cos^{-1}\left(\frac{s_i \cdot s_j}{|s_i| \cdot |s_j|}\right) \tag{3.1}$$

Here, $s_i = \begin{pmatrix} \cos\alpha_i \cos\delta_i \\ \sin\alpha_i \cos\delta_i \\ \sin\delta_i \end{pmatrix}$, and $s_j = \begin{pmatrix} \cos\alpha_j \cos\delta_j \\ \sin\alpha_j \cos\delta_j \\ \sin\delta_j \end{pmatrix}$. The two are the direction vectors for the guide stars *i* and *j*, respectively.

Similarly, denoting the star spot coordinates on the image plane of measured stars 1 and 2 as (X_1, Y_1) and (X_2, Y_2), respectively, then the angular distance in the star sensor coordinate system can be defined as follows (c.f. Fig. 3.1b):

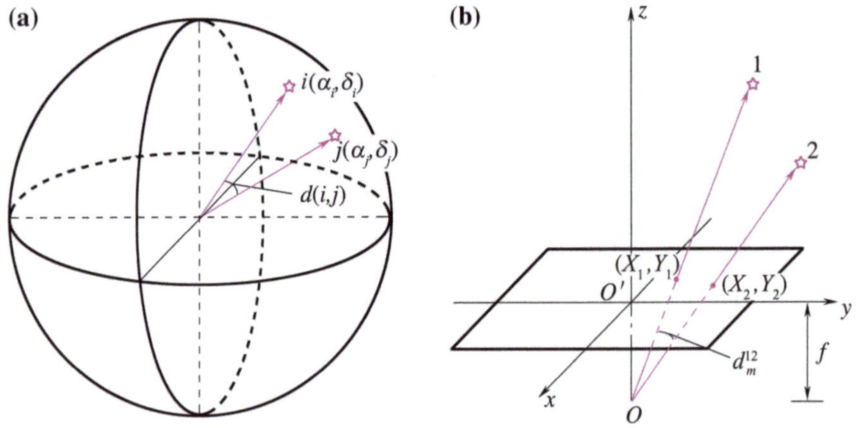

Fig. 3.1 Angular distance matching. **a** Angular distance in the celestial coordinate system. **b** Angular distance in the star sensor coordinate system

$$d_m^{12} = \cos^{-1}\left(\frac{s_1 \cdot s_2}{|s_1| \cdot |s_2|}\right) \tag{3.2}$$

Here, $s_1 = \frac{1}{\sqrt{X_1^2 + Y_1^2 + f^2}}\begin{pmatrix} X_1 \\ Y_1 \\ -f \end{pmatrix}$ and $s_2 = \frac{1}{\sqrt{X_2^2 + Y_2^2 + f^2}}\begin{pmatrix} X_2 \\ Y_2 \\ -f \end{pmatrix}$.

The two are the direction vectors of the measured stars 1 and 2 in the star sensor coordinate system, respectively.

If the measured stars can be matched with the guide stars, then

$$\left|d(i,j) - d_m^{12}\right| \leq \varepsilon \tag{3.3}$$

Here, ε refers to the error tolerance of angular distance measurement. For a measured star pair, there is generally more than one guide star pair that can meet the requirement mentioned above. Take the guide star database of stars of 6.0 Mv for example. If $d_m^{12} = 6°$, $\varepsilon = 0.02°$, then around 200 star pairs can meet the requirement in formula (3.3). Moreover, directional judgment in angular distance matching is also an issue, i.e., distinguishing the stars in the star pair from each other. To do this, only the value of angular distance is not enough, obviously. Other information, like star magnitude, must be used. Therefore, star pair matching through angular distance cannot fulfill the full-sky star identification independently.

Triangle matching is realized on the basis of star pair matching. With the addition of one more star, redundant star pairings can be eliminated to a great extent using the angular distances of the three sides. Triangle algorithms have two forms: "side-side-side" mode and "side-angle-side" mode. The measured triangle and guide triangle can be matched only if the following requirements are met at the same time (Fig. 3.2):

$$\begin{aligned} &\left|d(i,j) - d_m^{12}\right| \leq \varepsilon \\ &\left|d(j,k) - d_m^{23}\right| \leq \varepsilon \\ &\left|d(i,k) - d_m^{13}\right| \leq \varepsilon \quad \text{or} \quad \left|\theta(i,k) - \theta_m^{13}\right| \leq \varepsilon \end{aligned} \tag{3.4}$$

The major differences regarding those triangle algorithms lie in the selection and storage mode of guide triangles. The earliest research in this field was done by Liebe whose method stored all guide triangles that could be formed by guide stars

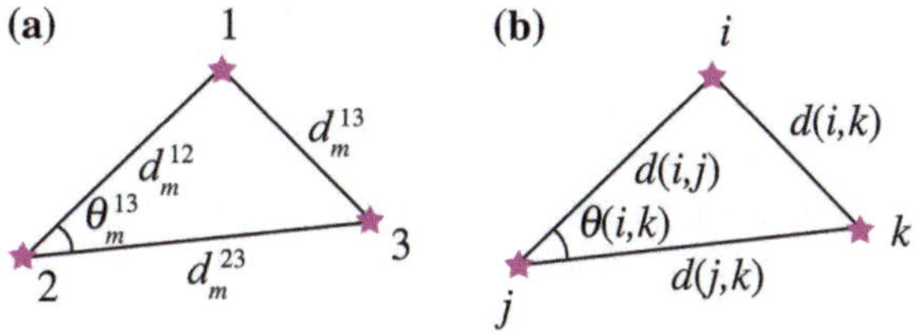

Fig. 3.2 Triangle matching. **a** Measured triangle. **b** Guide triangle

Table 3.1 Storage mode of guide triangles

$d(i,j)$ (°)	$d(j,k)$ (°)	$\theta(i,k)$ (°)
5.596	8.191	164.204
5.596	8.191	30.680
5.596	10.102	7.452
5.596	12.306	58.521
5.596	12.754	13.597
8.191	8.329	165.117
8.191	10.102	156.662
8.191	12.306	137.275
8.191	12.754	150.607
8.329	10.102	38.222
8.329	12.306	27.841
8.329	12.754	44.276

for the retrieval of matches. If one guide triangle and one measured triangle meet the requirement in formula (3.4), then the guide triangle is the match of the measured triangle. If there is only one guide triangle that is matched, then the star identification is successful. The guide triangles are stored in the ascending order of their first side $d(i,j)$ and the second one $d(j,k)$, as shown in Table 3.1, which indicates the storage mode of the guide triangles (the "side-angle-side" mode is used here). Liebe stores the angular distance values of each side (angle) of those guide triangles by sections, with error tolerance added, for quick retrieval. This triangle algorithm selects 1000 stars from 8000 guide stars to form guide triangles, but still requires around 1 MB memory to store 185,000 guide triangles. The identification rate using this method is 94.6%, which drops significantly (to 70–80%) when there are interfering stars. It takes 10 s for full-sky star identification on average.

3.1.2 Problems with the Triangle Algorithm

The major problem with the triangle algorithms is that there are too many guide triangles. Theoretically speaking, N guide stars can form $N(N-1)(N-2)/6$ guide triangles. Though the limitation by the FOV helps eliminate many of them, the number of the triangles left is still extremely large. This is particularly true when limiting magnitude is relatively high and the total number of guide stars comparatively large. Too many guide triangles will impede the use of the triangle algorithms. Therefore, triangle algorithms have to deal with the problem brought about by too many guide triangles. To solve this problem, guide stars and guide triangles need to be selected.

A modified triangle algorithm, proposed by Quine and Durrant-Whyte [10], holds that guide triangles contain a tremendous amount of redundant information, and thus one triangle for one star is enough. The principle of selecting triangles is as follows: use guide star S_1 as the first star of the triangle, then choose the brightest and the second brightest stars, S_2 and S_3 in the neighboring areas with a small radius r and a larger one R as the other two stars of the triangle to form a guide triangle (as shown in Fig. 3.3).

This method also holds true for measured stars in order to form measured triangles. With this method, N guide stars only need N guide triangles. The required memory capacity is also reduced. Though this method makes some progress in terms of time for identification and memory requirement, it still has problems:

① Measured triangles near the edge of the FOV may be selected in an erroneous way and then lead to a mistaken identification;
② It requires relatively accurate information on brightness, which is generally hard to obtain, when there may be some errors in the measurement of magnitude.

Kruijff et al. [11] propose a Douma/DUDE (Delta-Utec Douma Extension) algorithm based on Liebe's and Quine's triangle algorithms. Just like Quine's algorithm, the Douma/DUDE algorithm uses the information of brightness/magnitude, and selects triangles that are most likely to be detected as the guide triangles. The algorithm assigns each star a probability value which is related to the measured magnitude. The stars with the biggest probability value are most likely to be selected. Similarly, the triangles that are most likely to appear are selected preferentially. When making up measured triangles, Douma/DUDE also considers how much the location of measured stars (close to the edge of the FOV or not), will have an effect on identification. The Douma/DUDE triangle algorithm is more

Fig. 3.3 Quine triangle

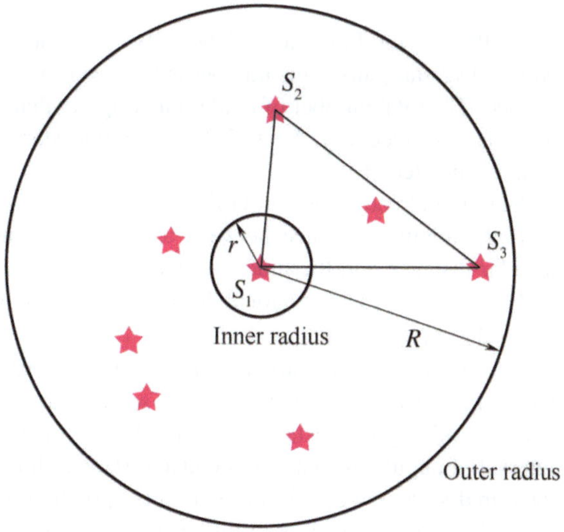

reasonable in the selection of guide triangles than the former two. However, it is also limited in practical use because relatively accurate information on brightness is still required.

3.2 Modified Triangle Algorithm Utilizing the Angular Distance Matching Method

As stated in the previous section, storing directly all guide triangles will lead to problems, such as tremendous memory requirement, many redundant matches, and too much time taken on retrieval of matches. However, the selection of triangles proposed in Quine's algorithm and Douma/DUDE algorithm also results in a higher probability of error in identification. Meanwhile, the requirements for the information on brightness are relatively strict. Regarding the above-mentioned problems, Zhang et al. [6, 7] propose a modified triangle star identification algorithm by using angular distance matching. The algorithm stores the data of angular distance of star pairs to realize the matching of triangles. The number of star pairs is much lower than that of triangles, so the memory capacity required decreases remarkably. At the same time, storing star pairs in a reasonable way helps speed up data retrieval and star identification. In this section, the specific realization process of the algorithm is introduced in detail, and an in-depth comparison of it with the traditional triangle algorithms is made.

3.2.1 *Star Pairs Generation and Storage*

Regardless of the limitation of the FOV, N guide stars can theoretically make up $N(N-1)/2$ star pairs. This number is far lower than that of the triangles that can be formed. The total number of guide stars brighter than 6.0 Mv is 5103, among which 3360 are selected (c.f. Sect. 2.2.1) as guide stars. The generation of star pairs follows the steps below.

Scan the already selected GSC.

If the angular distance between a star pair is smaller than d, then record the angular distance and the index numbers of the two stars, or the star pair.

Here, d refers to the diagonal distance of an FOV. For example, when the FOV is $12° \times 12°$, $d = 12\sqrt{2}°$.

Star pairs should be stored in the ascending order of the value of angular distance. In order to make it easier to search the matching of star pair's angular distance, all angular distances are divided into many intervals. The value of each interval of angular distance λ is equal to 0.02°. Thus, if the angular distance of two measured stars is worked out, their corresponding interval can be easily searched. The selection of star pairs in this interval will help to find out potential matching guide star pairs.

Fig. 3.4 Storage structure of star pairs

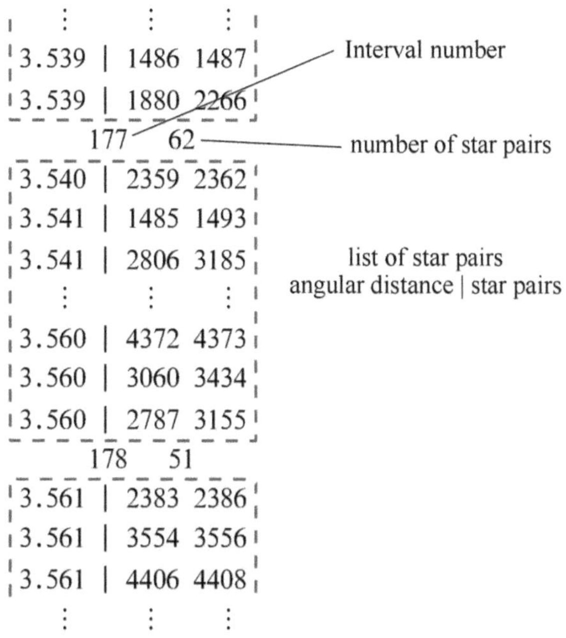

Figure 3.4 illustrates the storage of star pairs in interval No. 177, where all of the star pairs with an angular distance between 3.54° and 3.56° are stored. If the angular distance of two measured stars d is found to be 3.548, all the 62 star pairs in interval No. 177 can be selected as the probable matches of that measured star pair. If a higher error tolerance ε is used, then the star pairs in the neighboring intervals of No. 177, like No. 176 and No. 178, can also be the possible matches.

The curve in Fig. 3.5 shows how the number of star pairs in star pair database changes along with the change in angular distances. It can be seen that there is a linear co-relation: the bigger the angular distance, the higher the number of star pairs in the corresponding section. Figure 3.6 indicates the statistical probability of the distribution of the number of star pairs in different intervals of angular distance in the FOV of star sensors. It is obvious from Fig. 3.6 that angular distances of star pairs are mainly from 0° to 12°. When the angular distance is bigger than 12°, though there are a large number of star pairs in each interval of the database, these star pairs are less likely to appear in the FOV due to the limitation of its scope. Therefore, it is reasonable to think there are a large number of redundant star pairs in this interval that seldom show up in the real measured star image. Based on this, star pairs with angular distances of 0°–12° can be selected when a database of star pairs is established, and the angular distance values of the corresponding sides should also be in that range while selecting stars to form measured triangles. The method can largely reduce the size of the database of star pairs on the premise that the rate of identification is not affected.

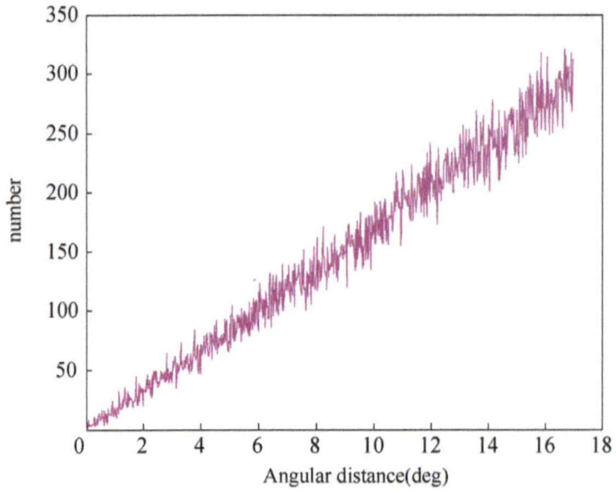

Fig. 3.5 The number of star pairs in star pair database under different values of angular distance

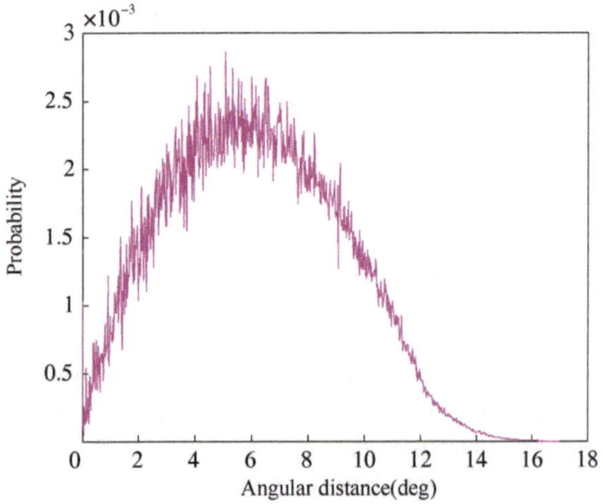

Fig. 3.6 Statistical probability of the distribution of the number of star pairs in different intervals of angular distance in the FOV

When $d = 12\sqrt{2}^{\circ}$, 122,964 star pairs are stored, and 1 MB memory is required if per star pair is calculated as 8 bytes. When $d = 12^{\circ}$, 60,692 star pairs are stored, and 0.5 MB memory is required. It is thus clear that a smaller value of d requires smaller memory. Experiments show that the identification rate will not be affected when $d = 12^{\circ}$.

3.2.2 Selection of Measured Triangles

Brighter stars are generally selected for star identification, as information provided by those stars is more reliable. During the selection of measured triangles, the principle similar to Quine's modified triangle algorithm and Douma/DUDE is adopted, that is, the brightest stars in the measured star image are selected preferentially to form measured triangles. The difference is that not just one measured triangle is selected for identification, but a group of bright stars (N_B in number) are selected to form measured triangles in a random manner. If one triangle fails in identification, the rest can be used until correct identification is obtained. Thus, this method only needs rough information regarding brightness, rather than fairly precise values which are used for a strict comparison of brightness.

The value of N_B can neither be set too large nor too small. If too large, there will be so many measured triangles that perhaps too much time will be needed in match retrieval. If too small, however, the risk of the inability to identify may be increased. Experiments demonstrate that $N_B = 6$ is a proper choice, that is to say, the top six brightest stars in the measured star image are selected to form measured triangles. If there are fewer than six measured stars, select all of them. The number of triangles that can be formed with six stars is $C_6^3 = 20$. Figure 3.7 illustrates the selection of measured triangles in a randomly generated star image. The bigger ○ indicates the selected measured stars which are used to make up measured triangles.

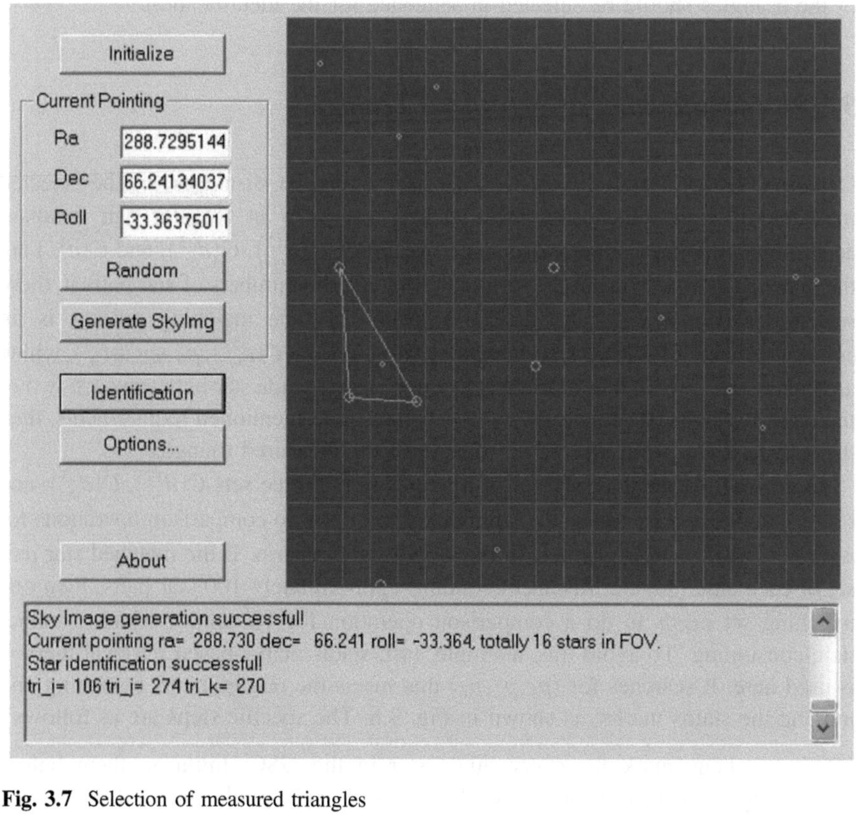

Fig. 3.7 Selection of measured triangles

For 20 possible measured triangles, if one side of any triangle among them has an angular distance larger than 12°, then ignore that triangle. For the rest of the triangles, the principles of selection are as follows:

① Give preference to the measured triangles formed by the brightest stars. Assuming M_1, M_2 and M_3 (M_i is measured by the gray value of star spot, the smaller M_i is, the brighter the star is) stand for the brightness of the three stars making up the triangle. Set $M = M_1M_2M_3$, the triangle with the smallest value of M is selected preferentially.

② Give preference to measured triangles with relatively shorter angular distances. It is obvious from Fig. 3.5 that a larger angular distance means a larger number of star pairs in the corresponding range and a higher probability of redundant matches. Therefore, among triangles with roughly equal M value, the ones with the shorter angular distance should be given priority.

Here, the principle of "the longest side the shortest" is adopted, that is, the triangle whose longest side has the shortest angular distance is selected preferentially.

Sort the measured triangles by following the above-mentioned selection principles. Those who rank higher are used for star identification successively. Once correct identification is obtained, next triangles can be skipped. Otherwise, the rest of the triangles should be selected in sequence for the identification.

3.2.3 *Identification of Measured Triangles*

Denoting d_m^{12}, d_m^{23}, and d_m^{13} as the three sides (angular distances) of the selected measured triangles, then find the matched star pairs in the star pair database according to the three angular distances. Assuming $C(d_m^{12})$, $C(d_m^{23})$, and $C(d_m^{13})$ are the matched star pair sets in the formula (3.3), and the numbers of star pairs in these sets are $n(d_m^{12})$, $n(d_m^{23})$, and $n(d_m^{13})$, respectively. The matching process is, in essence, to search for three star pairs $p_1 \in C(d_m^{12}), p_2 \in C(d_m^{23}), p_3 \in C(d_m^{13})$, which are linked end-to-end, that is, there is just one shared guide star between each two of the three star pairs. If set (p_1, p_2, p_3) meets the above-mentioned requirements, then the star pairs can form a matched triangle of the measured triangle.

In general, the match retrieval of triangles in the three sets $C(d_m^{12})$, $C(d_m^{23})$, and $C(d_m^{13})$, conducted by traversal combination, needs to do comparison operations for $n(d_m^{12}) \times n(d_m^{23}) \times n(d_m^{13})$ times in the worst case scenario. If the matched star pair set of each side (angular distance) contains approximately 100 star pairs, then one matching set needs to do a comparison operation for 10^6 times, which is rather time-consuming. To avoid this, a simple, fast, status marks-based retrieval method is used here. It searches for (p_1, p_2, p_3) that meets the requirements by setting and judging the status marks, as shown in Fig. 3.8. The specific steps are as follows:

① Set a status mark for every guide star in the GSC. Initialize them before matching and identification. Set the status of every guide star as 0.

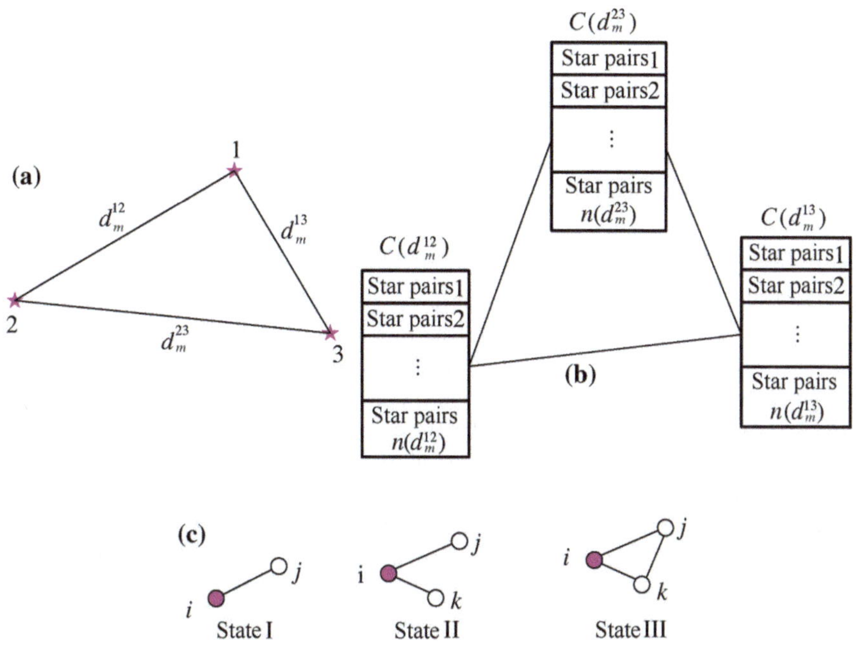

Fig. 3.8 The identification process of triangle. **a** Measured triangle. **b** Angular distance matching (star pairs). **c** State marks (I, II, III)

② Scan $C(d_m^{12})$. Set the status of the guide stars contained in all the star pairs in the set as I, and record the index number *j* of the other star in the same star pair (*i*, *j*).

③ Scan $C(d_m^{23})$. If the status of the guide stars contained in the star pairs in the set is already set as I, then set the status of the guide star as II, and record the index number of the other guide star k which constitutes a star pair with the former guide stars.

④ Scan $C(d_m^{13})$. If $(j,k) \in C(d_m^{13})$, which means the star pair formed by guide stars j and k is contained in the set $C(d_m^{13})$, then a matched triangle is successfully found. At the time, set the status of another guide star i, which forms star pairs with (j,k), from II to III. Thus, (i,j,k) is the guide triangle that matches with the measured triangle.

If the method of setting status marks is adopted, then the retrieval needs only $n(d_m^{12}) + n(d_m^{23}) + n(d_m^{13})$ times, far fewer than those of traversal combination.

3.2.4 *Process of Verification*

The foregoing identification process may bring about more than one set of guide triangles that can be matched with the measured triangle, thus other methods should be used to conduct further election. Verification is then introduced for this purpose. The main purposes of the verification are as follows:

① It can evaluate whether the identification is correct, and meanwhile rule out wrong matches, if any.
② It can identify the matched stars for as many measured stars in the measured star image as is possible, which is also beneficial to tracking identification and improving the accuracy of attitude calculation.
③ It can provide rough information on attitude.

The basic idea of verification is as follows. If the identification is correct, i.e., the guide triangle is the correct match of the measured triangle, then the attitude calculated with the result of the match must also be accurate. So the simulated star image (called reference star image) generated according to this attitude must be consistent with the original measured star image (i.e., the positions of star spots are identical). Figure 3.9 is an illustration of the correspondence of the guide stars distributed in the celestial sphere and the measured stars in the star sensor's FOV. When the identification is correct, then not only will the guide triangles correspond to the matched measured triangles, but there is a one-to-one correspondence between the other measured stars in the FOV and the guide stars in the celestial area.

Here, the calculation of attitude adopts the method of obtaining attitude with two vectors, that is, calculating the attitude of star sensor with the vector information regarding two matched stars. Its specific process is introduced in Sect. 1.2.1. This method is easy and fast, yet cannot generate a fairly precise attitude evaluation.

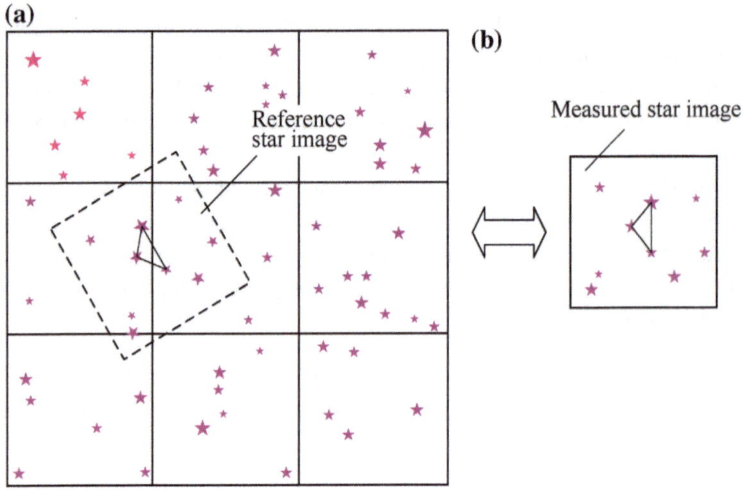

Fig. 3.9 Correspondence between guide stars and measured stars. **a** Guide stars in the celestial sphere. **b** Measured stars in the star sensor's FOV

Verification does not require precise attitude information, but needs fast attitude calculation. Therefore, the method of obtaining attitude with two vectors can meet the requirements here for verification.

The simulation process of the reference star image is described in Sect. 2.3. Based on the calculated attitude and the imaging parameters of the star sensor, the star image that the star sensor can capture with that attitude can be easily deduced. In order to do less calculations, the simulation of the reference star image here just involves getting the coordinate values of imaging star spots without involving the simulation of energy distribution and the synthesis of digital images and without considering the factors that can be ignored, such as distortion. The information in the partition table of GSC (c.f. Sect. 2.1) can be used for the fast retrieval of other guide stars in the neighborhood of the guide triangle that matches with the measured triangle.

Assuming Star 1 and Star 2 are two measured stars and their identification results are i and j (index number in the GSC), respectively. The generation process of the star image is illustrated in Fig. 3.10.

① Find in the GSC the index numbers of celestial area sub blocks, sub_i and sub_j, where the guide stars i and j are located respectively.

② In the partition table of GSC, find the record entries of the sub blocks sub_i and sub_j, and obtain the index numbers of the neighboring sub blocks. Record the sets of 3 × 3 sub blocks centering around sub_i and sub_j as $C(\text{sub}_i)$ and $C(\text{sub}_j)$, respectively.

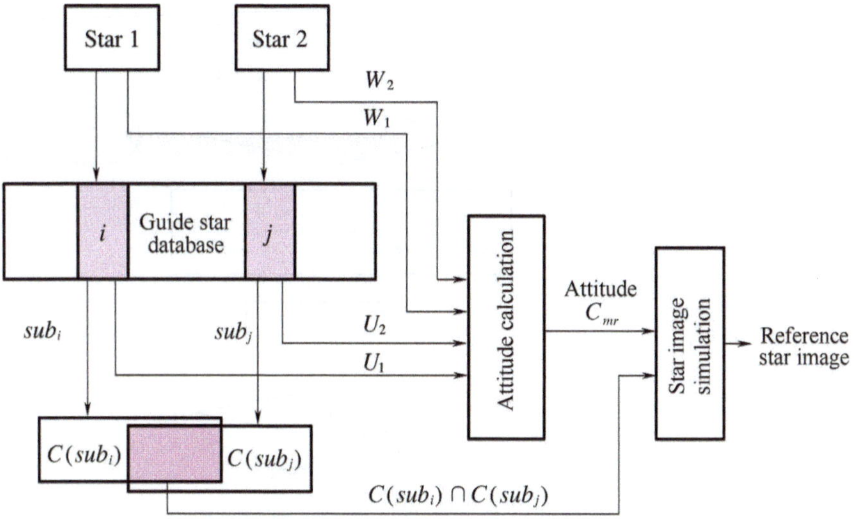

Fig. 3.10 Fast generation of star image

③ If the guide stars i and j are the correct identifications of the measured Star 1 and Star 2, then the reference star image (Fig. 4.9) must be located within both $C(\text{sub}_i)$ and $C(\text{sub}_j)$, i.e., the intersection $C(\text{sub}_i) \cap C(\text{sub}_j)$. Therefore, guide stars in the intersection area are used to generate the reference star image. With this method, the retrieval scope of guide stars is further narrowed down, which speeds up the generation of the star image.

If the stars (or the majority of them) in the reference star image can find their correspondent measured stars within the neighboring area of a small radius in the measured star image, then the reference star image and the measured star image are considered in correspondence. The identification is successful and the algorithm is finished. Otherwise, use the same method to verify other matched guide triangles. If the generated reference star image and the measured star image are not in correspondence, then the identification of measured triangles is not successful. Repeat the same identification process with the rest of the measured triangles of higher ranking. If there is no measured triangle which can be correctly identified, then the identification algorithm fails.

The flow chart of the whole algorithm is shown in Fig. 3.11.

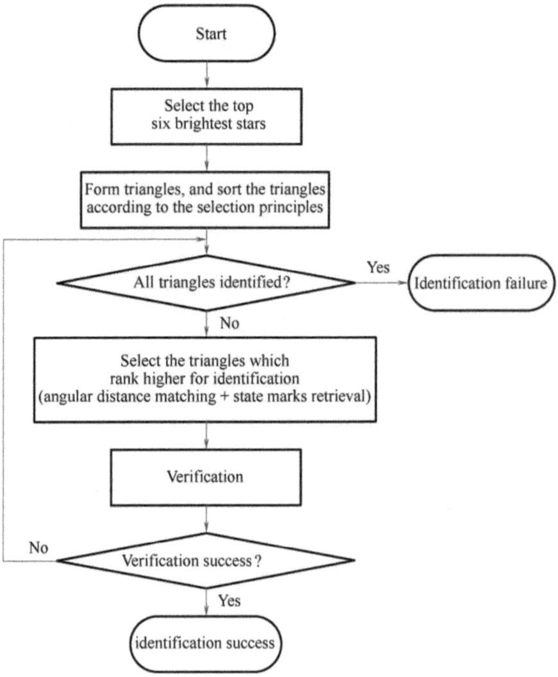

Fig. 3.11 The identification flowchart of the modified triangle algorithm

3.2.5 Simulations and Results Analysis

To evaluate the performance of the modified triangle algorithm, simulation experiments are carried out on star positional noise, magnitude/brightness noise, and the interfering star's impact on identification. The three noises or interferences are the main factors that affect star identification during the generation process of the star image. Before that, it is necessary to analyze the influence of the identification parameter N_S on identification rate and time.

1. Selection of Identification Parameter

N_S is the number of measured triangles used for identification. Theoretically speaking, a bigger N_S means a higher identification rate. When N_S is the biggest number of the available measured triangles combinations, the identification rate reaches the highest level. However, a bigger N_S may also mean too much time is needed for identification. This is particularly true when the uncertainty value ε of angular distance is rather high. An unsuccessful identification will be rather time-costly.

Figure 3.12a shows the results of random identification for 1000 times when N_S is at its biggest. In the experiment, the standard deviation of star positional noise is 0.5 pixel, magnitude noise 0.5 Mv, with no interfering star. The abscissa in Fig. 3.12a means the number of measured triangles that are used when the identification is correct (or wrong), and the ordinate means the number of times that this happens in the 1000 identification processes. It can be seen from Fig. 3.12a that in about 95% of all the cases, use of the first measured triangle can bring about a correct identification. The probability rises to more than 99% when the first three triangles are used.

Figure 3.12b indicates the average identification time under various conditions in the same experiment as shown in Fig. 3.12a. Denoting the average time of the first measured triangle in obtaining successful identification as 1, this is used as the benchmark of the identification time under other conditions. As the number of measured triangles being used increases, the identification time increases drastically. Unsuccessful identification takes the most time. By synthesizing the experiments results of identification rate and identification time, it can be analyzed that, during the algorithm identification process, Ns can be relatively small, not necessarily the same as the biggest number of measured triangles used for identification. Simulation experiments demonstrate that when $Ns = 3$, it can not only meet the requirements of the identification rate, but effectively reduces the average identification time as well. Therefore, $Ns = 3$ in all of the following simulations.

If the influences of interfering stars are also taken into account, when Ns is small, it is better to try to avoid the repeated use of measured stars contained in the Ns measured triangles. Its main purpose is, when bright stars (planets of high brightness, like Venus, Jupiter, etc.) become interfering stars, to avoid the cases when identification fails if all of the Ns measured triangles contain interfering stars.

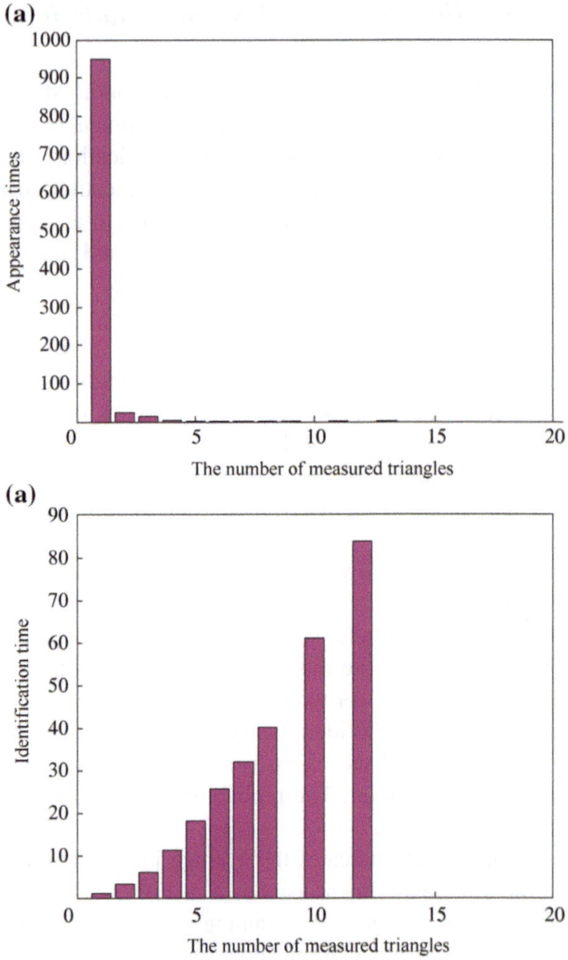

Fig. 3.12 Influences on the identification rate and identification time

2. Influence of Star Positional Noise on Identification

Star positional noise reflects the error of star spot centroiding, influenced by factors such as image noise, distortion of optical system, truncation error of sampling and quantitative, and the centroiding algorithm itself. To examine the identification algorithms' robustness to star spot centroiding error, a relatively big positional noise is generally used.

In experiments, for the simulated star image, a Gauss noise with mean = 0 and std. dev σ = 0–2 pixels, is added to the real star spot position. Select 1000 star images randomly from the celestial sphere, add positional noise according to the above-mentioned method, and then analyze the identification results. It is obvious from formula (3.3) that the value of ε varies with the level of noise. The value of ε in the triangle algorithms is determined by the interval λ during the storage process of angular distances. Two star pair databases when $\lambda = 0.02°$ and $\lambda = 0.05°$ are used for identification. The identification results are shown in Fig. 3.13.

It can be seen from Fig. 3.13 that the identification rates of different λ values are about the same under low level of noise situations. When σ is bigger than 0.5 pixel, the identification rate drops drastically when $\lambda = 0.02°$. When $\sigma = 2$ pixels, the identification rate is already below 84%. Under the same condition, the identification rate decreases relatively slowly when $\lambda = 0.05°$, and remains above 97% when $\sigma = 2$ pixels. The reason for a sharp decline in identification rate when $\lambda = 0.02°$ is that the uncertainty value ε of angular distances is so low that, when noise is added, the angular distance of a measured star deviates from the right matching interval and causes mistaken matches. A bigger λ can ensure that the angular distance between measured stars is always within the right matching interval after the noise is added. Meanwhile, it is noticeable that a bigger λ results in more star pairs to be searched for in angular distance matching, and more time in identification. At the same time, a bigger λ means larger memory usage in algorithm operation. Therefore, the value of λ is related to the level of noise, meaning a small λ should be taken when noise level is low, and the value of λ can be increased when the noise level is relatively high.

At low noise level ($\sigma < 0.5$), the modified triangle algorithm can obtain a nearly 100% identification rate, which is a remarkable improvement when compared with the 94.6% identification rate of Liebe's triangle algorithm. When $\sigma = 2$ pixels, the identification rate of Liebe's triangle algorithm drops rapidly to about 70%, while that of the modified triangle algorithm after increasing the value of λ still stands at 97%, or even higher.

3. Influence of Magnitude Noise on Identification

Magnitude noise reflects the precision of a photoelectric detector when measuring stars' brightness and is influenced by factors such as characteristics of the stellar

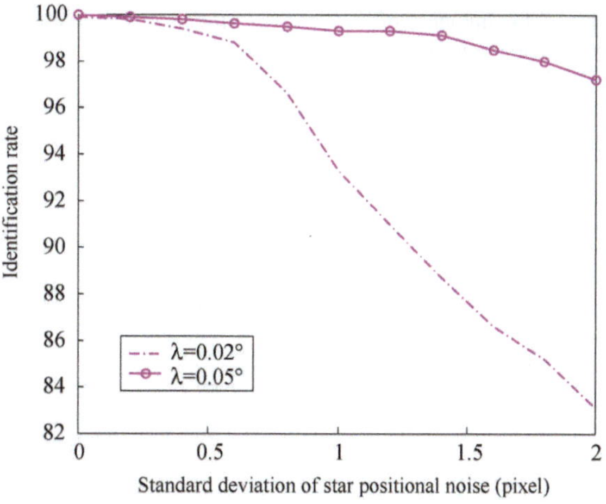

Fig. 3.13 Influence of star positional noise on identification

spectrum, characteristics of changes in star's brightness, imaging sensor, optical systems, and so on. The triangle algorithms use the rough information on brightness, for example, when selecting 6 stars from the measured star image, and when sorting the measured triangles that can be formed by the six stars. Therefore, it is necessary to assess the influence of magnitude noise on identification.

In experiments, during the simulation of a star image, a Gauss noise with mean = 0 and std. dev = 0–1 Mv is added to the magnitude, which is reflected in star image as noise of the gray amplitude of star spot images. 1000 star images are selected randomly from the whole celestial sphere for identification. Figure 3.14 shows the identification results.

It is clear from Fig. 3.14 that the modified triangle algorithm has a strong robustness to magnitude noise. When the standard deviation of magnitude noise is 1 Mv, the identification rate is still as high as 93%, and when the standard deviation of magnitude noise is less than 0.5 Mv, the identification rate maintains nearly 100%.

4. Influence of Interfering Stars on Identification

There are two types of interfering stars. The first are "artificial stars," like planets, nebula dust, space debris, and so on. It is hard to distinguish their imaging targets from ordinary star spot targets in measured star images. Moreover, due to the imaging sensor's limited detection capability of magnitude, some stars with lower brightness may also be captured. But they cannot capture their matched guide stars. The others are "missing stars," that is, the stars are supposed to be captured, but do not show up in the measured FOV due to particular reasons. Here, simulation experiments of the two situations are carried out to analyze their influences on identification rate.

Add to the measured star image a certain number of (one or two) artificial stars with equivalent magnitude ranging from 3 to 6 Mv and the standard deviation of

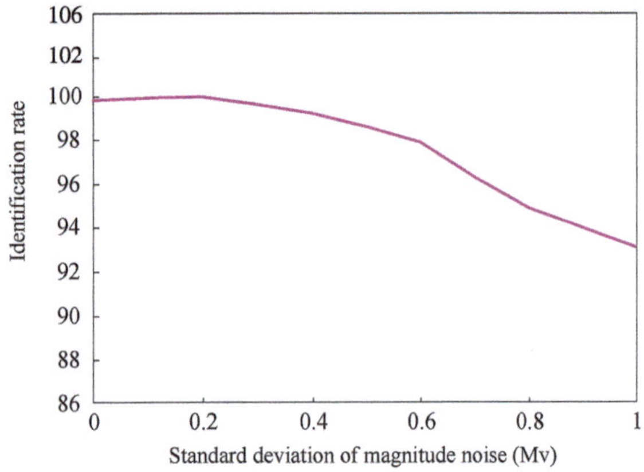

Fig. 3.14 Influence of magnitude noise one identification

magnitude noise being 0.2 Mv. The identification result is shown in Fig. 3.15. It is obvious from Fig. 3.15 that, when the equivalent brightness of the "artificial star" is relatively weak (>5 Mv, the triangle algorithm can ensure a comparatively high identification rate (>95%). When the equivalent magnitude is smaller than 5 Mv, the identification rate drops sharply, and when the equivalent magnitude is 3 Mv and there are two "artificial stars," the identification rate drops to around 50%. Other triangle algorithms, if interfered by bright artificial stars, will demonstrate a noticeable decrease in their identification rates. A bigger N_s can increase identification rate, but will lead to an increase in identification time at the same time.

Similarly, delete a certain number of (one to two) measured stars from the measured star image in order to examine the influence of "missing stars" on identification rate. The magnitude of deleted measured stars ranges from 3 to 6 Mv. The result is illustrated in Fig. 3.16. It can be seen that "missing stars" exert a comparatively small influence on identification.

5. Memory and Identification Time

The identification time of the algorithm is related to the interval parameter λ during the storage of angular distances. When $\lambda = 0.02°$, the average identification time is 8.4 ms, and when $\lambda = 0.05°$, the average identification time is around 10.3 ms. The identification time is the average of 1000 identifications operated on Pentium 800 MHz PC, and the codes have not gone through optimization. By contrast, the average full-sky identification time needed by Liebe's triangle algorithm is 10 s. Regardless of differences in hardware, the modified triangle algorithm is superior in terms of operation time.

The memory requirement of a modified triangle algorithm decreases largely due to the adoption of angular distance matching. For the selected 3360 guide stars, the

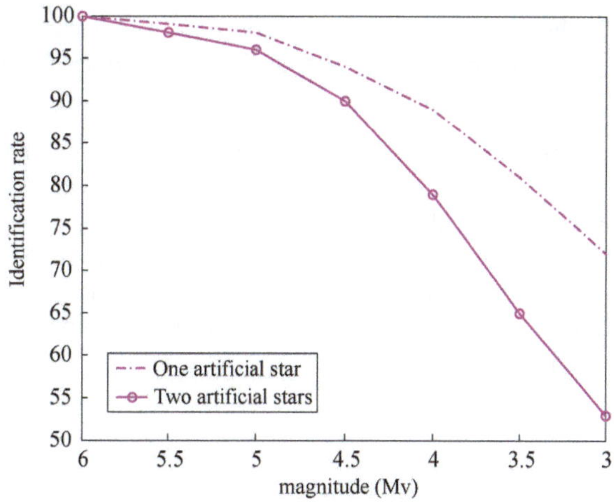

Fig. 3.15 Influence of "artificial stars" on identification rate

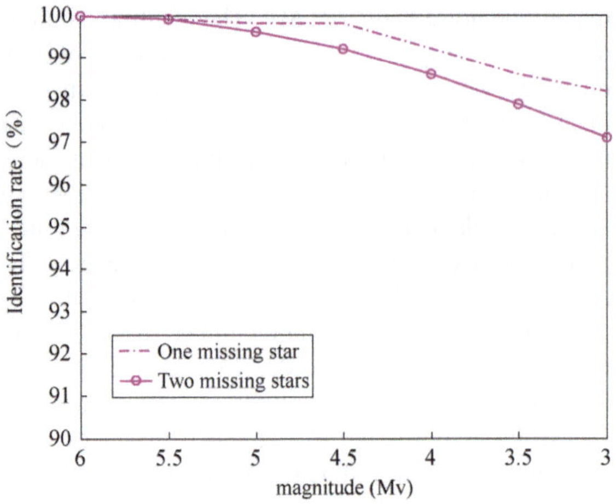

Fig. 3.16 Influence of "missing stars" on identification rate

memory required for storing angular distances is 0.5 MB, while the traditional triangle algorithm requires, in general, at least 1 MB.

3.3 Modified Triangle Algorithm Utilizing the *P* Vector

In application, the triangle algorithms need to compare the length of three sides (angular distances), thus the comparison times will multiply when the number of triangles is huge. Besides, a huge number of triangles make it difficult for storage and data retrieval. Generally, the following ways can be used to decrease the number of triangles:

① limiting the number of guide stars, and decreasing GSC capacity;
② storing only a part of triangles with certain characteristics;
③ optimizing the storage structure of triangles.

The above-mentioned methods all require comparison with measured triangles for multiple times, so their efficiency is low. To solve this problem, in this chapter, a modified star identification algorithm by using *P* vector is introduced. The algorithm selects the feature triangles of each star as the research subjects, calculates their feature values, and searches for their matched guide triangles through the feature values. It reduces the number of comparisons effectively and speeds up identification.

To reduce the number of triangles and the comparisons of triangles, the modified star identification algorithm by using P vector [8, 9] adopts Quine's principle that one triangle is for one star, and integrates the three sides of the feature triangles for each guide star into one parameter P. Compare the values of P to judge if the triangle is the matched one. The values of P make full use of the three sides, and reflect the features of triangles, so there is a one-to-one correspondence between the value of P and triangles, and fast star identification can be realized. Moreover, the verification procedure is added after the initial matching ends, to ensure that measured stars near the edge of the FOV, even though they cannot form feature triangles, can still be identified correctly.

3.3.1 *P* Vector Generation

The generation of P vector is divided into two steps: forming feature triangles and figuring out the optimal projection axis. The two steps are introduced below.

1. Forming Feature Triangles

Every guide star can form only one triangle, named as—feature triangle, with the guide star as the primary star. A feature star is made up of the primary star and two neighboring stars which are the closest to the primary star. The selection principle of neighboring stars is as follows: in the area between a small radius r and a larger one R, select the two stars which are closest and second closest to the primary star as the neighboring stars. The distance between the neighboring stars and the primary star must be shorter than the FOV radius R to ensure that those stars can be seen in the star sensor's FOV at the same time; and the distance must be longer than r to avoid the linking between the primary star and the neighboring stars during imaging. Therefore, the neighboring stars for forming feature triangles must meet the requirement $r < d < R$.

Feature triangles are spherical triangles on the celestial sphere. The side length is the angular distance between two stars. In the two sides linked to the primary star, if the rotation from the short side to the long side is counter-clockwise, θ is defined as positive. Otherwise, it is negative.

The three sides of a feature triangle formed according to the rules mentioned above make up a three-dimensional vector. When θ is positive, the three coordinates of the vector are all positive. Otherwise, they are negative. In Fig. 3.17, the distance between the primary star S_1 and the neighboring star S_2 is 3.654°. The distance between S_1 and the neighboring star S_3 is 5.864°. And the distance between S_2 and S_3 is 4.012°. The distance between S_1 and S_2 is smaller than that between S_1 and S_3, and the rotation is counter-clockwise, so θ is positive, and the three-dimensional vector corresponding to the feature triangle is (3.654, 5.864, 4.012).

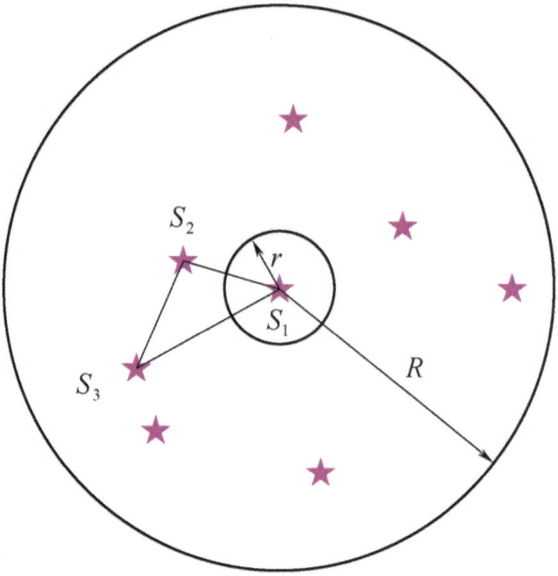

Fig. 3.17 Forming feature triangle

2. Figuring out Optimal Projection Axis

For quick retrieval of feature triangles, data needs to be organized in a certain way. A three-dimensional vector can be conceived as a point in the three-dimensional space. All vectors made up of feature triangles constitute a point set. If those points are scattered without overlapping when they are projected into a straight line, then only the feature triangle can be searched for through the projection points. Principal component analysis (PCA) is used to figure out the optimal projection principal axis. This method can reduce the data dimension. Therefore, the projection points have the best ability, and the similarities between the projection points have minimum changes, meaning that the interrelations of original points can be judged by observing the relative positions of projection points.

Suppose there are N feature triangles, then there are correspondingly N dimensional vectors. Denoting the direction of projection straight line as $\Omega = (\omega_1 \quad \omega_2 \quad \omega_3)^{\mathrm{T}}$, a certain dimensional vector as $X_i = (x_i \quad y_i \quad z_i)^{\mathrm{T}}$, then the coordinates of the projection point can be defined as Eq. (3.6) as follows:

$$P_i = \omega_1 x_i + \omega_2 y_i + \omega_3 z_i = \Omega^{\mathrm{T}} X_i \tag{3.6}$$

Thus, the mean and variance of all projection points are as follows:

$$\bar{P} = \frac{1}{N}\sum_{i=1}^{N} P_i = \frac{1}{N}\sum_{i=1}^{N} \Omega^{\mathrm{T}} X_i \tag{3.7}$$

$$
\begin{aligned}
D(P) &= \frac{1}{N}\sum_{i=1}^{N}(P_i - \bar{P})^2 \\
&= \frac{1}{N}\sum_{i=1}^{N}(P_i^2 - 2P_i\bar{P} + \bar{P}^2) \\
&= \frac{1}{N}\sum_{i=1}^{N}P_i^2 - \frac{1}{N}\sum_{i=1}^{N}2P_i\bar{P} + \frac{1}{N}\sum_{i=1}^{N}\bar{P}^2 \\
&= \frac{1}{N}\sum_{i=1}^{N}(\Omega^{\mathrm{T}}X_i)^2 - 2\bar{P}\left(\frac{1}{N}\sum_{i=1}^{N}P_i\right) + \bar{P}^2 \\
&= \frac{1}{N}\sum_{i=1}^{N}(\Omega^{\mathrm{T}}X_i)^2 - \bar{P}^2 \\
&= \frac{1}{N}\sum_{i=1}^{N}(\Omega^{\mathrm{T}}X_i \cdot X_i^{\mathrm{T}}\Omega) - \bar{P}^2 \\
&= \Omega^{\mathrm{T}}\left(\frac{1}{N}\sum_{i=1}^{N}X_iX_i^{\mathrm{T}}\right)\Omega - \bar{P}^2
\end{aligned}
\tag{3.8}
$$

Here, $i = 1, 2, 3, \ldots, N$. N stands for the number of projection points. The straight line in a certain direction, on which the projection points are discrete in the best way possible, is defined as the optimal projection principal axis. This means the variance of P_i in this case is the biggest. The key parameter for the optimal projection principle axis is the direction of the straight line, so a constraint of $\|\Omega\|^2 = \Omega^{\mathrm{T}}\Omega = 1$ can be added here to simplify the issue, without any effects on the results. Meanwhile, $\overline{P}^2$ is a constant, so the issue of figuring out the direction of the optimal projection principal axis can be turned into an optimization issue as follows:

$$
\begin{cases}
\max(D(P)) = \max(\Omega^{\mathrm{T}}Z\Omega) \\
\Omega^{\mathrm{T}}\Omega = 1
\end{cases}
\tag{3.9}
$$

Here, Z stands for the symmetric matrix $\frac{1}{N}\sum_{i=1}^{N}X_iX_i^{\mathrm{T}}$.

Defining Lagrange Function as follows:

$$
L(\Omega, \lambda) = \Omega^{\mathrm{T}}Z\Omega - \lambda(\Omega^{\mathrm{T}}\Omega - 1)
\tag{3.10}
$$

According to the concept of mathematical analysis, the necessary condition for the existence of extreme points is that

$$\begin{cases} \dfrac{\partial L(\Omega, \lambda)}{\partial \Omega} = 0 \\ \dfrac{\partial L(\Omega, \lambda)}{\partial \lambda} = 0 \end{cases} \tag{3.11}$$

i.e.

$$\begin{cases} 2Z\Omega - 2\lambda\Omega = 0 \\ \Omega^{\mathrm{T}}\Omega - 1 = 0 \end{cases} \tag{3.12}$$

$\Omega^{\mathrm{T}}\Omega - 1 = 0$ is obviously true. So Eq. (3.12) can be turned into Eq. (3.13):

$$Z\Omega = \lambda\Omega \tag{3.13}$$

λ and Ω are the eigenvalue and eigenvector, respectively, of the symmetric matrix Z. Then the objective function $\max(\Omega^{\mathrm{T}}Z\Omega)$ is as follows:

$$\max(\Omega^{\mathrm{T}}Z\Omega) = \max(\Omega^{\mathrm{T}}\lambda\Omega) = \max(\lambda\Omega^{\mathrm{T}}\Omega) = \max(\lambda) \tag{3.14}$$

It can be seen from Eq. (3.14) that the maximum value of the objective function $\max(\Omega^{\mathrm{T}}Z\Omega)$ is the maximum feature value of symmetric matrix Z. The optimal projection principal axis is thus the eigenvector corresponding to the maximum eigenvalue, i.e., the optimal projection direction of data points when they are projected from three-dimensional space to one-dimensional space.

When the optimal projection principal axis is figured out, its positional relationship with data point set is shown in Fig. 3.18. Verify the P values of these projection points. There are no identical values, that is, any one projection point is corresponding to only one three-dimensional vector, and a one-to-many correspondence does not exist. Even if some projection points are very close to one another, they will be considered as candidates during the identification process.

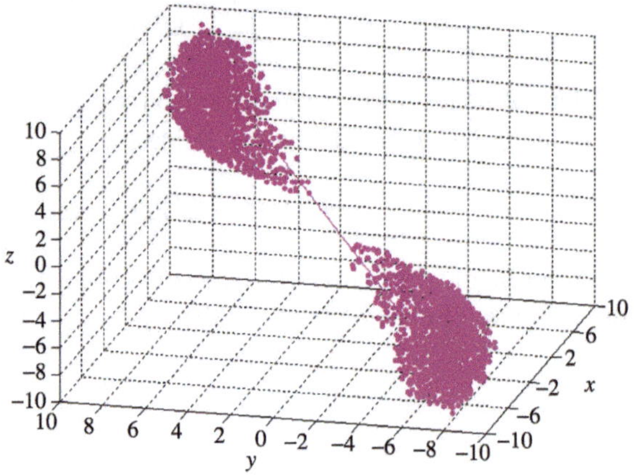

Fig. 3.18 The optimal projection principal axis of triangle

3.3.2 Construction of a Guide Database

Guide database is the star catalog and pattern information used when a star sensor identifies a star image and calculates attitude. A guide database consists of three parts: a GSC, a feature triangles database and a *P* value vector table. The GSC stores the position and magnitude of guide stars. The feature triangles database stores the basic information—vertex and side length—of feature triangles. The *P* value vector table stores the *P* values and the index number of every triangle. The construction of GSC is described in Chap. 2. Here, the construction of feature triangles database and the structure of *P* value vector database are mainly introduced.

1. Feature Triangles Database

Feature triangles are formed through using every star in GSC as a primary star. The feature triangle database stores the information of these triangles. The index number of each guide star, which is the vertex of the triangle, and the corresponding lengths of three sides, are recorded as one entry. All entries are stored based on the index number of primary stars. The storage structure of feature triangles database is illustrated in Fig. 3.19. The data is used to figure out the optimal projection principal axis, calculate *P* values of all the projection points, and search for guide stars that match the measured stars during identification.

2. Structure of *P* Value Vector Table

P values obtained from feature triangles are also stored in the guide database. Each *P* value has its corresponding feature triangle, so that each *P* value and the index number of the primary star of its feature triangle are stored within an entry. Thus the *P* value vector table is established. For efficient retrieval, all these entries are stored in an ascending order of *P* values. The storage structure of the *P* value vector table

i	*j*	*k*	short edge	long edge	third edge
107	102	103	3.500280	5.444810	3.253172
			…		
2344	2341	2338	2.882823	4.966379	2.517098
			…		
4235	4231	4228	1.998139	3.804716	4.896636
			…		

Fig. 3.19 Storage structure of feature triangles database. *i* index number of primary star S_1 in GSC, *j* index number of neighboring star S_2 which is close to the primary star, *k* Index number of neighboring star S_3 which is far from the primary star, *short edge* angular distance between S_1 and S_2, *long edge* angular distance between S_1 and S_3, *third edge* angular distance between S_2 and S_3

is illustrated in Fig. 3.20. Here, "*P*" stands for the value obtained from feature triangles based on Eq. (3.6), and "index" is the index number of the primary star of the corresponding triangle.

Once a *P* value is obtained, its corresponding feature triangle can be searched for quickly with the help of the *P* value vector table. Moreover, the direction vector of the optimal projection principal axis is also stored in the database.

3. Mapping Relation between the Two Databases

Mapping between the *P* value vector table and the feature triangle database is realized through the index number of primary stars. The mapping relation is illustrated in Fig. 3.21. After a P' value is obtained, its equivalent number in the *P* column of the *P* value vector table is found. According to the index number I' to the right of this value, the primary star whose entry number (i) is I' can be found in the feature triangle database. That entry is the feature triangle that corresponds to the P' value.

P	index
−9.784504	4235
−9.783472	2344
−9.771813	107
...	...

Fig. 3.20 Storage structure of *P* value vector table

P	index
−9.784504	4235
−9.783472	2344
−9.771813	107
...	...

i	*j*	*k*	short edge	long edge	third edge
107	102	103	3.500280	5.444810	3.253172
			...		
2344	2341	2338	2.882823	4.966379	2.517098
			...		
4235	4231	4228	1.998139	3.804716	4.896636
			...		

Fig. 3.21 The mapping relation between *P* value vector table and feature triangles database

3.3.3 Matching and Identification

The identification process of this algorithm has two parts: initial matching and verification of identification. Initial matching begins with the calculation of the only P value of a certain measured triangle, followed by a quick retrieval of the corresponding feature triangle on the basis of the P value. A successful identification of a triangle indicates that the orientation information of three stars has been obtained. Two of them can be chosen to work out the general attitude of star sensor at the time, and an ideal simulated star image can be acquired with the imaging model of the star sensor. The similarities between the measured star image and the simulated one can be compared in order to verify the results of identification.

1. Initial Matching

After a measured star image is captured, the first thing is to observe if there are three or more measured stars. With only one or two stars, identification is impossible. Otherwise, the algorithm in this section can be used. The process of initial matching is as follows:

1. Select the primary star. The star near the center of the FOV is preferred, while for those close to the edge of the FOV, chances are that not all the neighboring stars needed in forming the feature triangle are seen in the FOV.
2. Determine the neighboring stars. Sort the stars surrounding the primary star according to distance. Then calculate their angular distances with the primary star. Select the two stars closest to the primary star and in the area of $r < d < R$ as neighboring stars to form the feature triangle.
3. Figure out the P value. The three-dimensional vector $X = (x \quad y \quad z)$ which describes the features of the feature triangle can be worked out through its angular distances. The projection axis vector stored in the P value vector table is retrieved and then its projection on the optimal projection principal axis, i.e., the value of P, is obtained according to Eq. (3.6). Then the direction in which the short side can rotate to the long side is worked out to determine whether the P value is positive or negative.
4. Match triangles. If there is no error in the feature triangle of the measured star, then the P value vector table has the one and only P value which corresponds to the result in step (3). The corresponding feature triangle of the guide star can thus be determined. However, due to various errors, the projection points tend to deviate from the locations where they are supposed to be. As a result, triangles whose P values are within the error range of $[P - \varepsilon, \quad P + \varepsilon]$ are all considered as candidates to be examined. Then the three sides of the measured triangle and the candidate feature triangles of guide stars are compared respectively. If the errors of all the three sides are all very small, and only one triangle meets this

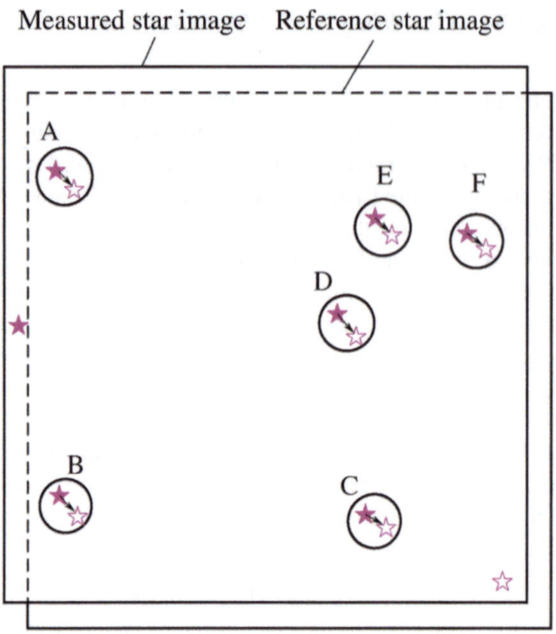

Fig. 3.22 Verification with reference star image

requirement, then this feature triangle is regarded as being the match of the measured triangle, with vertexes in a one-to-one matching relationship. However, if several matching triangles are found, then verification must be carried out for further selection.

2. Verification of Identification

This verification process is similar to the modified triangle algorithm method by using angular distance matching in that it generates a reference star image based on the result of initial matching and the imaging parameter of the star sensor. As shown in Fig. 3.22, "☆" stands for a star in the reference star image, and "★" for a star in the measured star image. The two star images may be not exactly the same when compared, yet if most stars in the reference star image have their corresponding measured stars in an area with a small radius in the measured star image, then the two star images are viewed as the same, that is, identification is successful. Nevertheless, if the two star images are different, which means the identification of feature triangles becomes wrongly calculated, then another primary star should be selected from the rest of the measured stars to form a new feature triangle for matching. If no feature triangles can get matched successfully, the identification algorithm fails.

The flow chart of the identification algorithm is shown in Fig. 3.23.

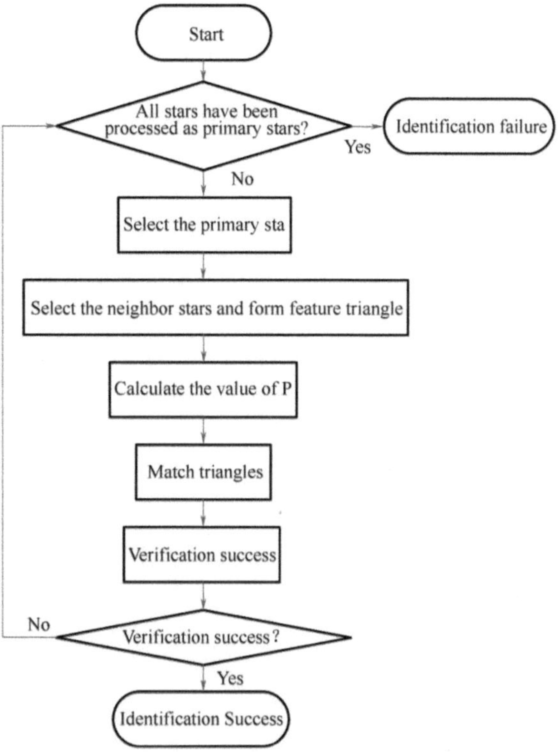

Fig. 3.23 Flowchart of the star identification algorithm by using the *P* vector

3.3.4 Simulations and Results Analysis

The guide star database used in the simulations is based on the SAO Star Catalog. Stars of brightness greater than 6 Mv (5103 in total) are chosen from the star catalog to constitute the guide star database. The size of the FOV is 10.8° × 10.8°, and the focal length of the lens is 80.047 mm. The pixel size is 0.015 mm, and the resolution is 1024 × 1024 pixels. The simulation processes are realized on Intel Pentium4 2.0 GHz computers.

1. Identification Example

Figure 3.24 indicates the results of identification of four randomly generated star images by using *P* vector identification algorithm. "+" which stands for a measured star in the FOV, and "o" for a correctly identified star. It is obvious that this algorithm makes a correct identification of all the measured stars in the simulated star images.

2. Influence of the Number of Measured Stars in the FOV on Identification Rate

The number of measured stars in the FOV is an essential factor that can influence the identification rate. If the number of measured stars in the FOV is too small, it is

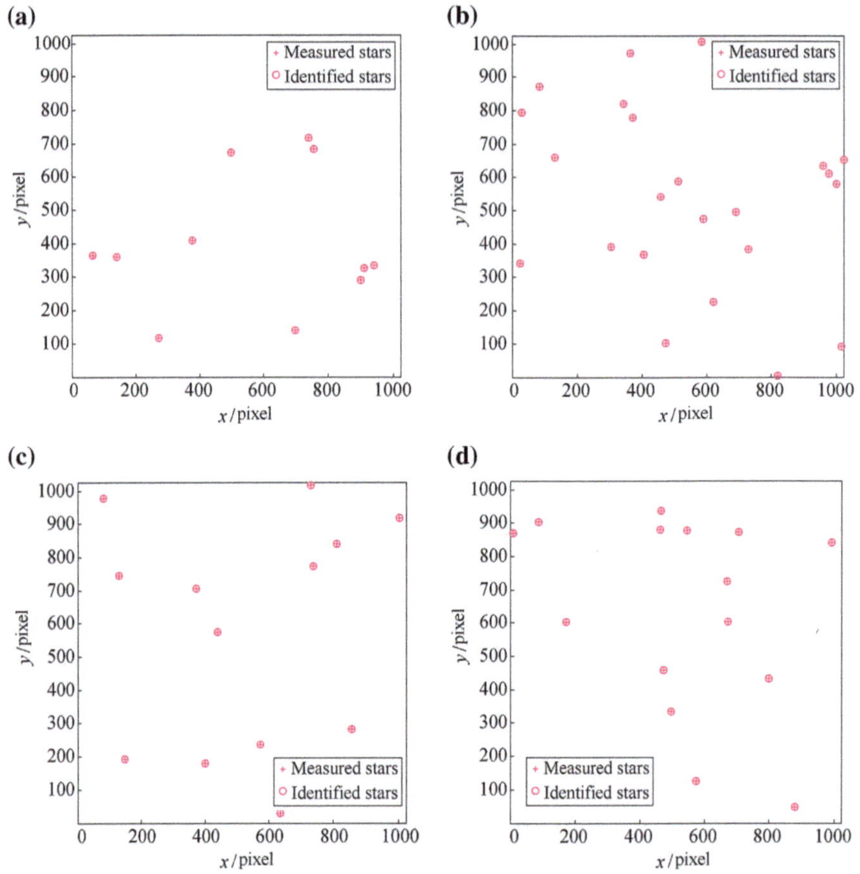

Fig. 3.24 Identification results of four randomly generated star images

hard to find more stars to verify the result of initial matching during the verification process. As a result, even if there are several candidate triangles in the process of initial matching, a final match cannot be identified. However, if there are enough measured stars in the FOV, this will not occur and a high identification rate can be guaranteed. It can be observed in Fig. 3.25 that with five measured stars in the FOV, the identification rate stands at merely 76%, which will further drop if the number is lower. But when there are more than six measured stars, the identification rate increases remarkably. When the number exceeds ten, the rate is close to a 100%.

3. Influence of Positional Noise on the Identification Rate

Similar to the foregoing simulations, noises are added to the coordinates of star spots in the reference star image to make comparisons with other algorithms for differences in identification results under the same conditions. To precisely evaluate

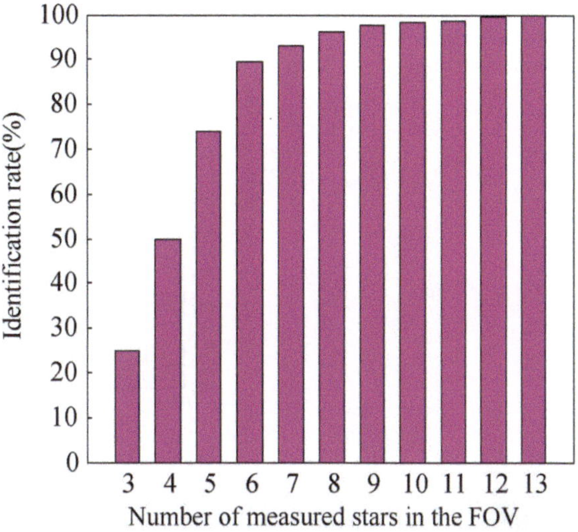

Fig. 3.25 Influence of the number of measured stars on identification rate

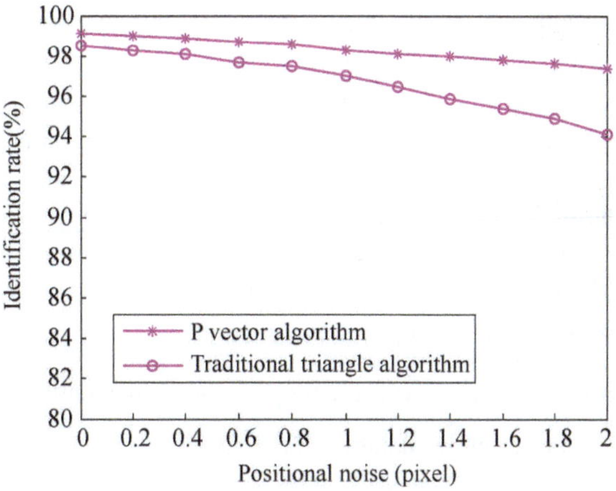

Fig. 3.26 Influence of positional noise on identification rate

the algorithm's performance of identification, the Monte Carlo method is adopted here to select randomly from the whole celestial sphere a 1000 star images for identification under every noise condition, and then the identification results can be obtained.

Figure 3.26 shows the curve of identification rates when the standard deviation of positional noise changes from 0 pixel to 2 pixels. It is obvious that the algorithm displays strong resistance to positional noise. Even when the standard deviation of positional noise is as high as 2 pixels, the identification rate of the algorithm still stands at 97%. The reason for the high identification rate lies in that, during initial matching, triangles within a certain error range of the *P* value are all regarded as candidates. When affected by noise, the projection value *P* of some triangles shifts,

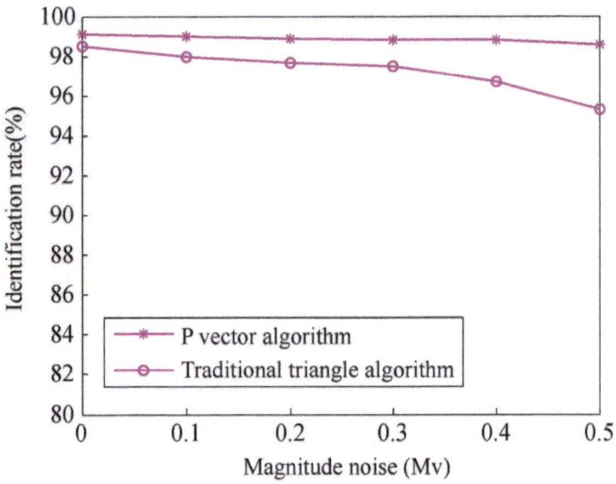

Fig. 3.27 Influence of magnitude noise on identification rate

so a relatively big tolerance should be adopted in comparing *P* values to ensure high identification rates.

4. Influence of Magnitude Noise on Identification Rate

Figure 3.27 shows the identification results by using the *P* vector star identification algorithm and the traditional triangle algorithms when a gauss noise with mean = 0 and std. dev σ = 0–0.5 Mv is added. The *P* vector algorithm does not use brightness information when forming triangles. So even when there is a huge magnitude noise, it does not affect the identification rate. In contrast, the traditional ways introduce magnitude information to reduce the number of triangles to be formed. Therefore, the identification rate drops when a greater magnitude noise is added.

5. Memory and Identification Time

Memory requirement for the *P* vector star identification algorithm is as follows:

Feature triangles database: 218 KB
P value vector table: 89 KB
GSC for storing guide stars: 179 KB
Information in the partition table: 57 KB

Thus, a total memory of 543 KB is required. After many simulations, the average time needed for identification operated on Intel Pentium4 2.0 GHz processor platform is 2.064 ms.

References

1. Liebe CC (1995) Star trackers for attitude determination. IEEE Trans Aerosp Electron Syst 10 (6):10–16
2. Liebe CC (1992) Pattern recognition of star constellations for spacecraft applications. IEEE AES Mag 28(6):34–41
3. Ju G, Kim H, Pollock T et al (1999) DIGSTAR: a low-cost micro star tracker. AIAA-99-4603
4. Scholl M (2019) Star field identification algorithm—performance verification using simulation star fields. SPIE 1993:275–290
5. Mortari D, Junkins J, Samaan M (2001) Lost-in-space pyramid algorithm for robust star pattern recognition. In: 24th annual AAS guidance and control conference, AAS 01-004
6. Wei X (2004) A research on star identification methods and relevant technologies in star sensor. Doctoral thesis of Beijing University Aeronautics and Astronautics, Beijing, pp 1–14
7. Zhang G, Wei X, Jiang J (2006) Star map identification based on a modified triangle algorithm. Acta Aeronautica Et Astronautica Sinica 27(6):1150–1154
8. Yang J (2007) A research on star identification algorithm and RISC technology application. Doctoral thesis of Beijing University Aeronautics and Astronautics, Beijing, pp 1–17
9. Yang J, Zhang G, Jiang J (2007) Fast star identification algorithm using P vector. Acta Aeronautica Et Astronautica Sinica 28(4):897–900
10. Quine BM, Durrant-Whyte HF (1996) Rapid star pattern identification. SPIE 2739:351–360
11. Kruijff M et al (2003) Star sensor algorithm application and spin-off. In: 54th international astronautical congress of the International Astronautical Federation (IAF), the International Academy of Astronautics and the International Institute of Space Law, vol 1, pp 349–359

Chapter 4
Star Identification Utilizing Star Patterns

Traditionally, most star identification algorithms are based on the characteristics of angular distance, such as polygon match algorithms, triangle algorithms and group match algorithms. Though they are easy to use, these algorithms require a relatively large storage capacity because their matching features are line segments (angular distances) or triangles. The Liebe triangle algorithm, for example, stores all guide triangles, and the storage requirement for only 1000 guide stars is as high as 1 MB. Another kind of star identification algorithm is utilized in order to use "star patterns", which regards the geometric distribution characteristics of stars (neighboring stars) as in the neighborhood of a measured star (or guide star) as its feature pattern. This feature pattern is the star's sole "signature", which distinguishes it from other stars. In this sense, this kind of star identification is closer to general pattern matching. Generally speaking, the algorithms by using "star patterns" have stronger fault-tolerant abilities and smaller storage needs. Recently, star identification methods have tended to adopt algorithms by using "star patterns", among which the grid algorithm is the most representative. Compared with traditional algorithms, the grid algorithm is outstanding in performance, but deficient in feature extraction.

In this chapter, three star identification algorithms by using star patterns are introduced, including star identification by using radial and cyclic star patterns, by using Log-Polar transformation, and star identification without calibration parameters. This chapter also introduces implementation of these algorithms in detail, and then compares their performances with that of the grid algorithms through simulations.

G. Zhang, *Star Identification*, DOI 10.1007/978-3-662-53783-1_4

4.1 Introduction to the Grid Algorithm

As the representative of star identification algorithms by using star patterns, the grid algorithm performs better in fault-tolerance, storage capacity and running time. This section presents a brief introduction to the principles and performance features of the grid algorithm, and analyzes its disadvantages in its feature extraction approach.

4.1.1 Principles of the Grid Algorithm

The grid algorithm was firstly proposed by Padgett [1], and it is a star identification method by using "star patterns". Its pattern generation process (as is shown in Fig. 1.17) can be roughly divided into the following steps:

① Determine the primary star (the star to be identified) and pattern radius *pr*. The pattern of the primary star consists of neighboring stars in the neighborhood determined by the *pr*.
② Shift the star image so that the primary star is the center of the FOV.
③ Within radius *pr* and beyond radius *br*, find the star l which is closest to s. l is called the location star.
④ With the line connecting the primary star and the location star as the initial coordinate axis and the primary star as the origin, rotate the star image and divide the image into a grid of $g \times g$. In this way, the primary star's feature pattern is expressed in the grid $cell(i,j)$. If there are neighboring stars in the grid cell, the corresponding value is 1, otherwise, it's 0.

Denoting it by the one-dimensional vector, and suppose the star's feature vector is

$$\begin{aligned} v &= (a_1, a_2, \ldots, a_k, \ldots, a_{g^2}), \\ k &= 1, 2, \ldots, g^2 \\ a_k &= \begin{cases} 1 & \text{cell}(i,j) = 1 \\ 0 & \text{cell}(i,j) = 0 \end{cases} \end{aligned} \tag{4.1}$$

Here, $k = j \cdot g + i$.

Denoting measured star j's pattern as pat_j, and the pattern set of all guide stars in the GSC as $\{\text{pat}_i\}$, in essence, star identification's aim is to seek

$$\max_i \text{match}(\text{pat}_j, \text{pat}_i) \tag{4.2}$$

Here, in $\text{match}(\text{pat}_j, \text{pat}_i) = \sum_{k-1}^{g^2} (\text{pat}_j(k) \,\&\, \text{pat}_i(k))$, and stands for logic and operation.

Table 4.1 Comparison of storage capacity

Algorithm	$\|C\| = 7548$	$\|C\| = 11901$
Triangle algorithm	13,400 triangles = 3 MB	555,000 triangles = 12 MB
Group match algorithm	66,000 pairs = 0.7 MB	166,000 pairs = 1.6 MB
Grid algorithm	7548 patterns = 0.5 MB	11901 patterns = 0.7 MB

Table 4.2 Comparison of average running time

Algorithm	$\|C\| = 7548$	$\|C\| = 11901$
Triangle algorithm	1.8 s	8.3 s
Group match algorithm	1.6 s	7.4 s
Grid algorithm	0.04 s	0.12 s

Compared with traditional algorithms, the grid algorithm is better in performance. Padgett has compared the performance of the grid algorithm, the group match algorithm and the triangle algorithm in detail [2]. The statistics of storage capacity and the average running time are shown in Tables 4.1 and 4.2, and $|C|$ here stands for the scale of the selected guide star catalog.

The average running time is measured on the same hardware platform. In the experiment, the standard deviation of star spot position noise is 0.5 pixel, and that of star magnitude noise is 0.3 Mv.

When the standard deviation of star spot position noise is 1 pixel, the identification rate of the grid algorithm is still close to 100%, while the rate of the group match algorithm and the triangle algorithm is only around 90%. When the standard deviation of noise rises to 2 pixels, the identification rate of the grid algorithm is about 95%, while that of the other two algorithms drops rapidly to 50–60%.

When the star magnitude noise's standard deviation is 0.5 Mv, the identification rate of the grid algorithm is close to 100%, while that of the group match algorithm and the triangle algorithm is only about 80%. When the star magnitude noise's standard deviation increases to 1 Mv, the identification rate of the grid algorithm can still be around 100%, while that of the other two algorithms drops sharply to 50–60%.

It is clear through these experimental data that the grid algorithm by using "star patterns" enjoys significant advantages compared with traditional algorithms based on angular distance.

4.1.2 *Deficiencies of the Grid Algorithm*

In spite of the grid algorithm's advantages shown above, it is deficient in feature extraction mainly in two aspects:

① The probability of the correct selection of the location star (location star) is low. The probability is only about 50% even if there is no star spot position noise.

And the rise of positional noise further decreases the probability. There are two major reasons: Firstly, the primary star is near the edge of the FOV, which makes the location star more likely to fall outside the FOV; Secondly, due to the errors in star magnitude measurement, accuracy of luminosity information cannot be guaranteed to be of a high enough accuracy. A mistaken determination of a location star generates an incorrect feature pattern, and if so, it's almost impossible to obtain a correct identification. To compensate for this possibly inaccurate determination of location star and the consequent misidentification rate, the grid algorithm increases the number of stars to be identified to ensure a still relatively high accuracy calculation. The grid algorithm adopts the FOV of $8° \times 8°$, and selects stars brighter than 7.5 Mv as guide stars. The total number of guide stars after selection is around 13,000. The average number of stars in a round FOV of radius 4° is close to 30. Therefore, a larger portion of stars could obtain a more accurate identification calculation even if other measured stars were identified incorrectly because of a potentially mistaken selection of location stars. The identification rate of the grid algorithm will drop significantly when the average number of stars in the FOV is low.

② The feature pattern cannot reflect the degree of internal similarity. The grid of $g = 8$ is shown in Fig. 4.1. Suppose its feature pattern vector is pat. According to the construction process of grid's feature pattern vector, the element at positions (14, 19, 39, 45, 51) in pat is 1, and the element at other positions is 0. Affected by errors in star spot position measurement, the star on the edge of the grid might move from place A to place B, and suppose the feature pattern vector extracted in this way is pat′. Apparently, the element at positions (19, 22, 39, 45, 51) in pat′ is 1, and the element at other positions is 0. It is thus evident

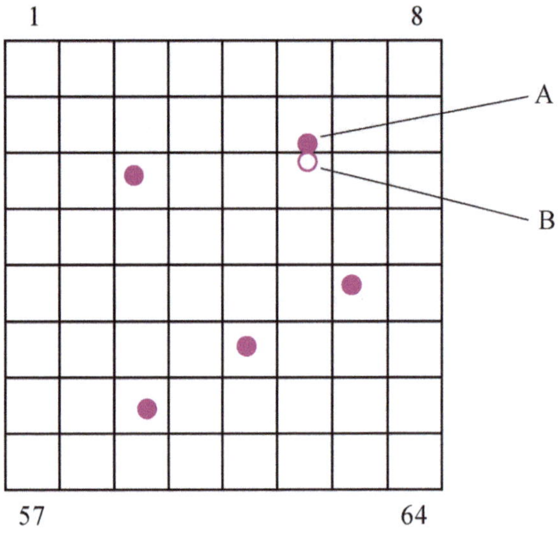

Fig. 4.1 A grid of 8×8

that, with this method, there exists big differences between feature vectors extracted, based on similar distribution features. In other words, feature vectors' similarity cannot be reflected in feature space.

4.2 Star Identification Utilizing Radial and Cyclic Star Patterns

To solve problems of the grid algorithm's feature extraction, Zhang et al. [3–5] have proposed star identification algorithms by using radial and cyclic star patterns. This section presents the detailed implementation of this algorithm, and compares its performance with that of the grid algorithm through simulations.

4.2.1 Star Patterns Generation and Storage

To avoid the grid algorithm's problems, neighboring stars' distribution features are resolved into radical and cyclic directions (Fig. 4.2). First, the rotation-invariant radial feature is reliable enough to be directly applied to matching and identification without the determination of location stars. Next, the similar features would stay similar after extraction in feature space because both radial and cyclic features are one-dimensional. The star identification method, by using radial and cyclic star patterns proposed here, is developed from this idea.

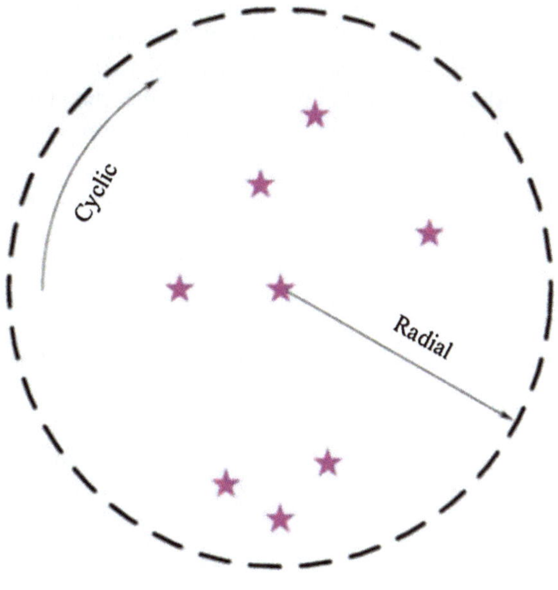

Fig. 4.2 Feature extraction in radial and cyclic way

Fig. 4.3 Radial feature

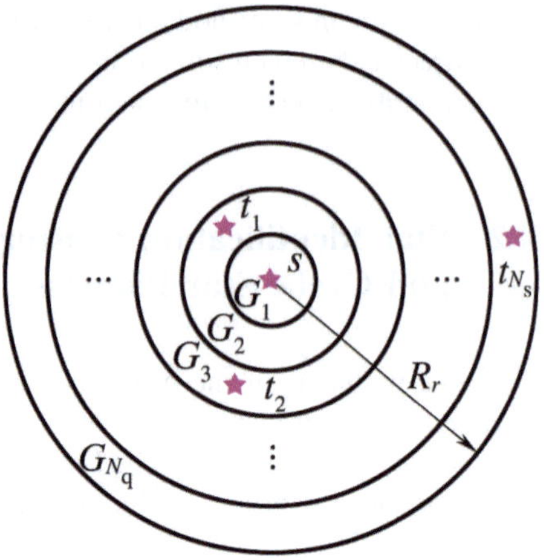

There are differences between radial and cyclic features. The radial feature enjoys rotation invariance, and it is a reliable characteristic, while the cyclic feature, like the grid algorithm, needs a location star to generate a cyclic feature pattern. Given this distinction, a multi-step match is adopted. The radial feature of star pattern is used for an initial match, and then a follow-up match is carried out using a cyclic feature of the star pattern. In initial match, guide stars are limited within a small range so that mistaken matching would be decreased as much as possible. The radial feature is just featured with this reliability and meets the demand. The cyclic feature can be used for a follow-up match to further eliminate redundant matches.

Construction process of radial feature is shown as follows (as is shown in Fig. 4.3):

① With s as the primary star, determining the radial pattern radius R_r. Stars in the neighborhood of the radius R_r are called neighboring stars of s. These neighboring stars constitute the radial pattern vector of s.

② Along the radial direction, the neighborhood of radius R_r which centers on s is divided into rings $G_1, G_2, \ldots, G_{Nq}$, intervals between which are equal. (Nq here means the grade of subdivision.)

③ Calculate the angular distance between neighboring star t_1 $(i = 1, 2, \ldots, Ns)$ and s and set it as $d(s, t_i)$, then this neighboring star i $(i = 1, 2, \ldots, Ns)$ falls in the ring $\mathrm{int}(d(s, t_i)/R_\mathrm{r})$ (int means rounding), thus the radial feature pattern vector of s is denoted as

$$\begin{aligned} &\mathrm{pat}_r(s) = (B_1, B_2, \ldots B_j, \ldots B_{Nq}) \\ &j = 1, 2, \ldots, Nq \end{aligned} \tag{4.3}$$

$$B_m = \begin{cases} 1 & G_m \text{ with neighboring star} \\ 0 & G_m \text{ without neighboring star} \end{cases}$$

Fig. 4.4 The cyclic feature

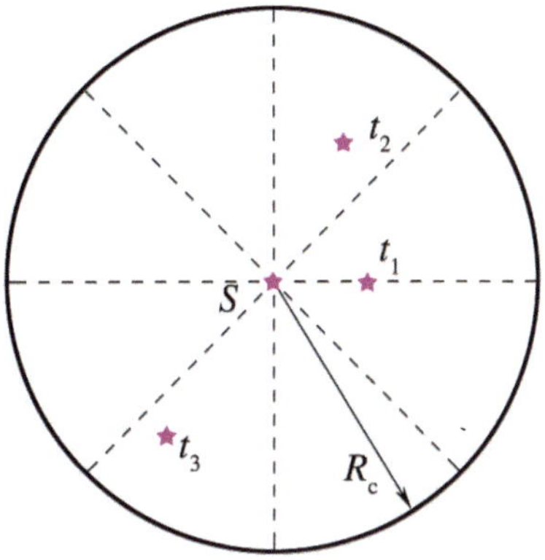

Take three neighboring stars, for example, to illustrate the construction process of the cyclic feature (Fig. 4.4). The steps are as follows:

① With s as the primary star, determine the cyclic pattern radius R_c and neighboring stars t_1, t_2, t_3.
② With s as the primary star and origin, calculate the angles between neighboring stars successively (see $\angle t_1 s t_2, \angle t_2 s t_3, \angle t_3 s t_1$, in Fig. 4.4).
③ Find the smallest angle ($\angle t_1 s t_2$) and choose the side of this angle (st_1) as the starting side to evenly partition the neighboring round area into eight sectors.
④ A vector v of eight bit is formed according to the neighboring stars' distribution in the sectors counter-clockwise. If there are neighboring stars in this sector, the corresponding bit is one. Otherwise, it is zero. In Fig. 4.4, it is shown that $v = (11000100)$.
⑤ Shift v circularly to find the maximum number (decimal) as the cyclic pattern of s. v remains unchanged after the shift as shown in Fig. 4.4, and the cyclic feature $\text{pat}_c(s) = (11000100) = 196$.

Under special circumstances, $\text{pat}_c(S) = 0$ when the number of stars in the neighborhood is 0, and when the number of stars in the neighborhood is 1, $\text{pat}_c(S) = 128$. The smallest angle between neighboring stars is found in a radial feature extraction, which is similar to the determination of the location star in the grid algorithm. Therefore, the cyclic feature is not reliable. However, it has little effect on the identification process because it's used only in the follow-up match.

Star identification is conducted after the radial and cyclic features of all guide stars are extracted utilizing the above-mentioned method. With the same method, features of the measured star image are extracted to find a guide star whose pattern is closest to that of the measured star by using the matching criterion similar to Eq. (4.2). However, a big problem of matching in this way is that it is too time-consuming. For the screened 3360 guide stars, it needs 672,000 (3360 × 200) comparison operations for one match of the radial feature if the radial grade of subdivision $Nq = 200$. Apparently, traversal search in this way takes too much

Record Index number of a guide star

⋮

50 | 33 34 117 118 250 251 354 455 457 480 481 633 768 769 …

49 | 1331 1334 1335 1429 1430 1939 1942 2021 2023 2161 2162 …

83 | 74 75 81 82 320 321 327 328 368 476 477 581 583 641 656 …

72 | 204 205 216 217 345 622 639 640 829 830 1332 1335 1337 …

84 | 23 24 139 140 296 297 373 374 930 931 1055 1056 1064 …

99 | 86 88 130 154 155 156 166 168 214 217 221 223 265 313 …

⋮

Fig. 4.5 Storage structure of LT with the radial pattern

time. To avoid this problem, a lookup table (LT) is designed to store guide stars' radial features, and matching speeds up significantly with this kind of storage structure.

The LT has Nq entries which are denoted respectively as $\mathrm{LT}_i(i = 1, 2, \ldots, Nq)$ and correspond to Nq rings in the radial feature extraction. Take each guide star in the GSC as the primary star and construct its radial feature pattern vector with the method introduced above. A new record is put in the LT_j if there is a neighboring star within ring G_i, and this record is the primary star's index number. After searching all neighboring stars of this primary star, the corresponding records are put in the LT_j. The LT is constructed after all guide stars in the GSC are searched.

Figure 4.5 shows a part (from the third entry to the eighth entry) of the LT with radial pattern vector $R_{\mathrm{r}} = 10°$ and subdivision grade $Nq = 200$. The LT's structure is very simple. Each entry only includes its record number and the guide stars' index numbers in ascending order. During the construction of LT, the guide star's index number which appears repeatedly in the same entry should be eliminated.

Because the radial feature is used for initial match and the cyclic feature is used for follow-up matches, the set of candidate stars is limited within a comparatively small range after the initial match. Therefore, the speed of a matching search is not a major problem anymore, and the cyclic feature can be stored directly for matching. The structure of the navigation database is shown in Fig. 4.6.

4.2.2 Process of Identification

In a multi-step match, an initial match is performed firstly with the radial feature so that the search range is limited within a relatively small order. Next, screening by other features layer by layer is conducted until the correct match is finally obtained. The identification rate of star identification by using "star patterns" is related to the neighboring stars' number n_{neighbor} in the neighborhood determined by the pattern

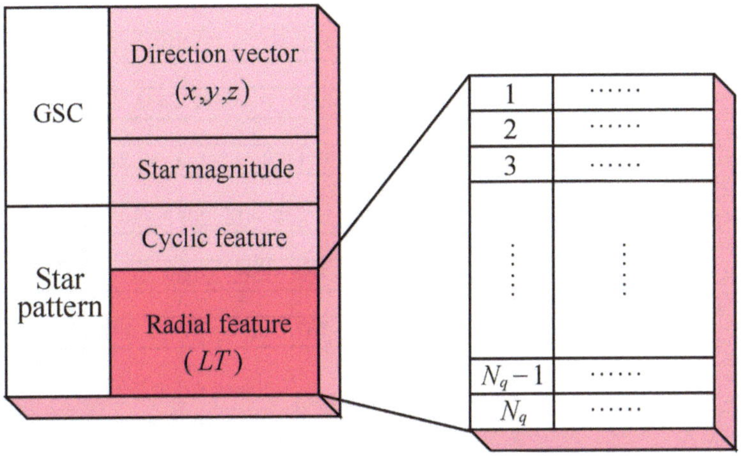

Fig. 4.6 Structure of the navigation database

radius. Generally speaking, the bigger n_{neighbor}, the higher the identification rate. Smaller n_{neighbor} provides little information with too many redundant matches and cannot guarantee correct identification. Therefore, a brighter measured star with a bigger n_{neighbor} value is the priority choice. Defining $Q = M/n_{\text{neighbor}}$ (M for the star magnitude), and ordering measured stars according to the value of Q. α stars of smallest Q are therefore selected successively for matching.

(1) Initial Match

The initial match is related to the structure of LT. Take a measured star s as an example to illustrate the process of the initial match. Suppose s's radial feature is denoted as (12, 26, 31, 54, 102, 133). That is, s has neighboring stars in rings (12, 26, 31, 54, 102, 133). Search the records in the (12, 26, 31, 54, 102, 133) entries of the LT for the index numbers of guide stars (Fig. 4.7). The guide star with the index number of 454 appears five times. Both of the guide stars whose index numbers are 2294 and 211 appear twice, respectively. And all of the other stars appear only once. It indicates that, for the guide star of index number 454 and the measured star s, there are five neighboring stars appearing in the same and corresponding rings. Therefore, this guide star is most likely to become the match star of s. The initial match with this method is described as follows:

Distribute N (N as the total number of chosen guide stars) counters $(CT_1, CT_2, \ldots, CT_N)$ to correspond to each guide star. Take the measured star to be identified in the measured star image as the primary star and construct its radial feature pattern vector by using the method described above. If there are neighboring stars in ring G_j, scan all records in the LT_j and the value of the counter of their corresponding guide stars' index number is added by 1. Finally, compare $(CT_1, CT_2, \ldots, CT_N)$ and choose a guide star with the highest counter value and this guide star is likely to be the match star (called the candidate star) of the measured

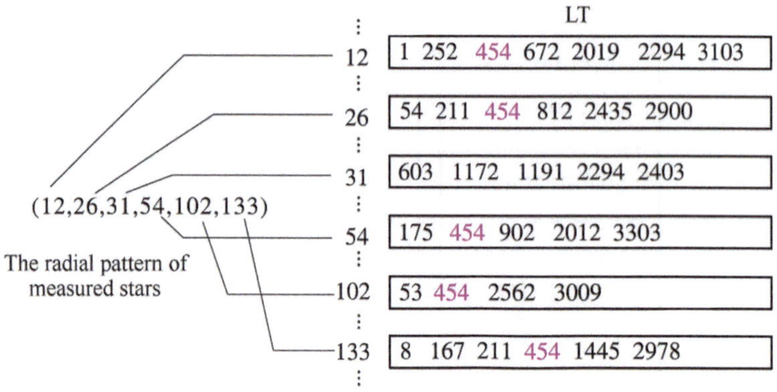

Fig. 4.7 Initial match

star. An initial match is conducted with α stars, selected from the measured star image, and their matching stars are found. The candidate star(s) that the measured star i corresponds to may not be unique, so the candidate stars' set is recorded as can_i. In essence, the initial match is to narrow down the scope of search matching from the whole guide star catalog to $\{\mathrm{can}_i\}$ $(i = 1, 2, \ldots, \alpha)$.

(2) Follow-up Match

In theory, if there are two or more measured stars, whose candidate star is unique after initial match, the next step goes directly to verification and identification. However, when the number of stars in the FOV is relatively small, there are a large number of redundant matches in $\{\mathrm{can}_i\}$. Here, the cyclic feature vector is used for further screening: if the measured stars' candidate star is not unique, the candidate star's cyclic feature vector is constructed using the above-mentioned method. And the candidate star's cyclic feature vector stored in the cyclic pattern database is compared with the constructed cyclic feature vector above. If the two vectors are the same, this candidate star is kept: otherwise, it's removed.

(3) The FOV Constraint

Under some circumstances, the candidate star obtained after screening with both radial and cyclic distribution features is still not unique. If so, further screening, based on other constraints must be carried out. It is shown in the experiment that all correct matches of the measured stars in the image cluster in a certain area of the FOV, while the incorrect matches (error and redundant matches) are randomly distributed across the sphere. The method of FOV constraint is based on this principle. If the number of stars in some candidate star's neighborhood of radius r is under a certain threshold value T, this candidate star is eliminated directly from the candidate star(s).

(4) Verification and Identification

For the candidate stars obtained after the above screening, verification and identification are conducted if there are two or more measured stars with a unique candidate star. Compute the attitude matrix of star sensor according to two measured stars and the direction vector of their matches in the star catalog. With this attitude matrix, generate a reference star image, which is similar to the generation of the star image. Compare the reference star image with the measured star image. If star positions in the measured star image correspond to those in the reference image, the match is correct and the identification is successful. Otherwise, continue the verification with other match stars. If the verifications always fail, the identification also fails. The concrete process and function of the verification are detailed in Sect. 3.2.4.

The flow chart of star identification by using radial and cyclic star patterns is shown in Fig. 4.8.

4.2.3 Simulations and Results Analysis

Parameters used in star sensor's imaging during simulations are the same as those used in simulations in Sect. 3.2.5. Simulations mainly include the selection of subdivision grades and radial pattern radius, the effect of star spot position noise, star magnitude noise and interfering star value on identification (compared with the grid algorithm), and the effect of the number of stars in the FOV on identification.

(1) Selection of Identification Parameters

Initial match is the most important step in identification algorithm, in which two key parameters are used: radial pattern radius R_r and radial quantizing grade Nq.

Figure 4.9 demonstrates how the identification rate varies along with the different radial pattern radiuses and radial quantizing grades. Here, the standard deviation of star positional noise is defined as a 0.5 pixel and the identification is conducted in 1000 random orientations in the celestial sphere. It is seen that the bigger R_r value, the higher the identification rate. But the increase in the identification rate is little when $R_r > 10°$. The selection of Nq is related to the noise level of the star position. The highest identification rate is achieved only when an appropriate Nq is chosen. When Nq is bigger, the quantizing grade is finer and the algorithm is easier to be interfered with by noise and more likely to provide a wrong match; A smaller Nq results in redundant matches easily. To enhance the algorithm's robustness to positional noise, a smaller Nq should be taken.

Figure 4.10 shows how LT capacity varies along with changes in radial pattern radius and quantizing grades. The ordinate stands for the total number of storage records in the LT. It is shown that the required storage capacity increases quickly with the increase of R_r, and Nq has little influence on the storage capacity.

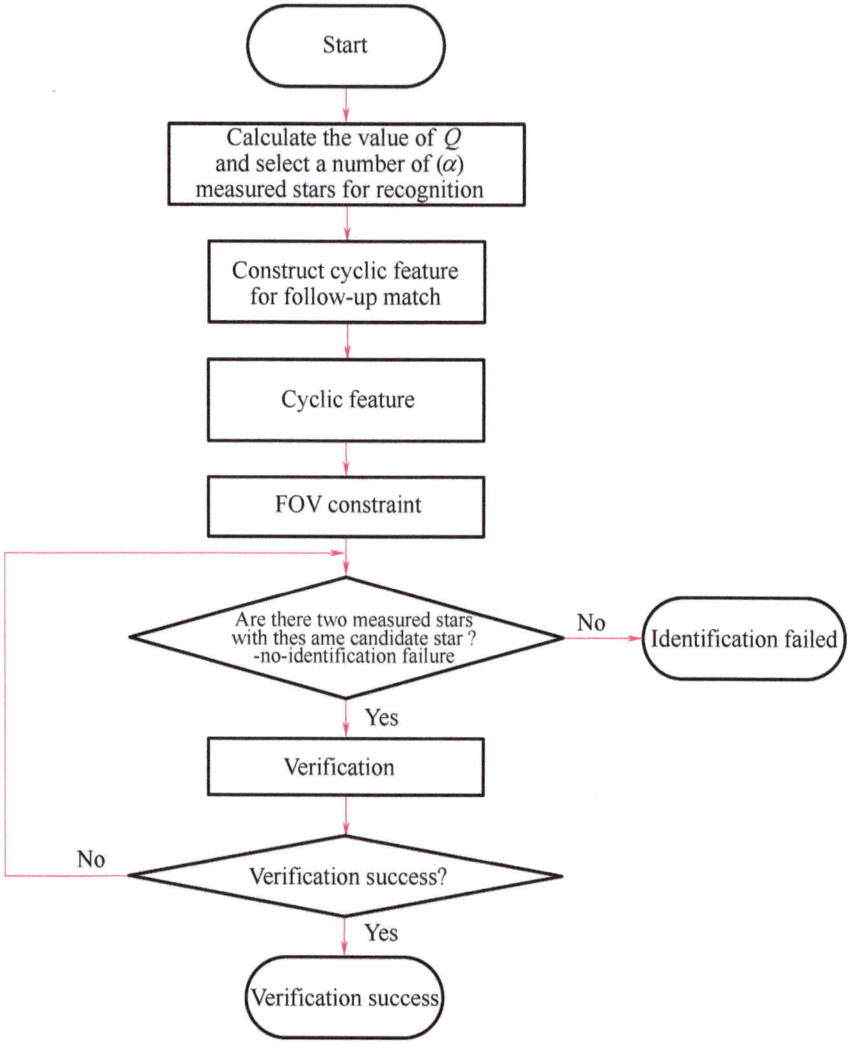

Fig. 4.8 Flow chart of the star identification algorithm by using radial and cyclic star patterns

To ensure that algorithm's identification rate is high enough and the required storage capacity is as small as possible, the parameters of initial match R_r and Nq are defined as 10° and 200, respectively. In addition, here the cyclic pattern radius $R_p = 6°$ during the construction of the cyclic feature.

(2) Effect of Star Spot Position Noise on Identification

To investigate the influence of star location error on algorithm's identification rate, a gauss noise with mean = 0 and std. dev $\sigma = 0 - 2$ pixels is added to the true star position in star image generated through simulation. Figure 4.11 shows the

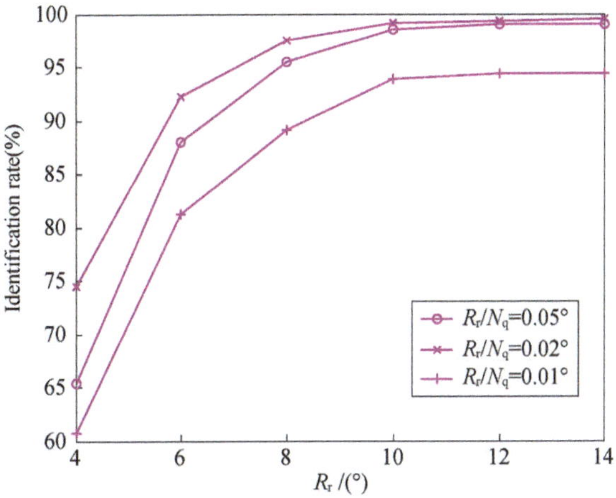

Fig. 4.9 Effect of radial pattern radius and quantizing grade on identification rate

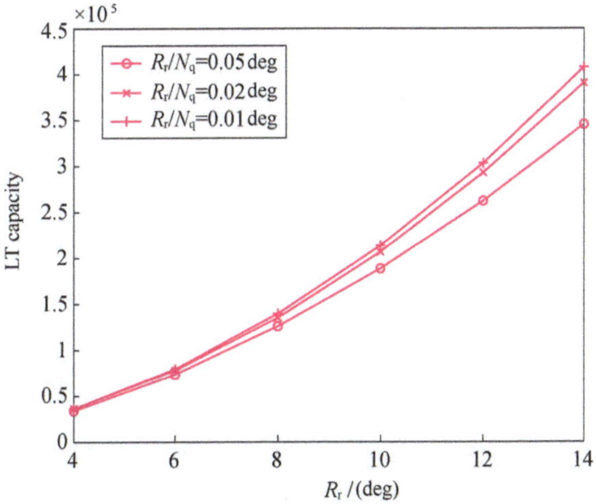

Fig. 4.10 Effect of radial pattern radius and quantizing grade on LT's capacity

statistics of identification results of 1000 star images randomly chosen from the celestial sphere. Comparing these results with those of the grid algorithm in the same experimental conditions. With the grid algorithm, pattern radius $R_p = 6°$ and the number of grid cells $g^2 = 60 \times 60$. It's shown in Fig. 4.11 that this algorithm always performs better than the grid algorithm in identification rate when star spot position noise changes. When the standard deviation of position noise is 2 pixels, the identification rate of this algorithm is about 97%, while that of the grid algorithm drops to around 94%.

(3) Effect of Star Magnitude Noise on Identification

To investigate the effect of brightness error on identification, a gauss noise with mean = 0 and std. dev = 0−1 Mv is added to the star magnitude in the star image

Fig. 4.11 Effect of position noise error on identification rate

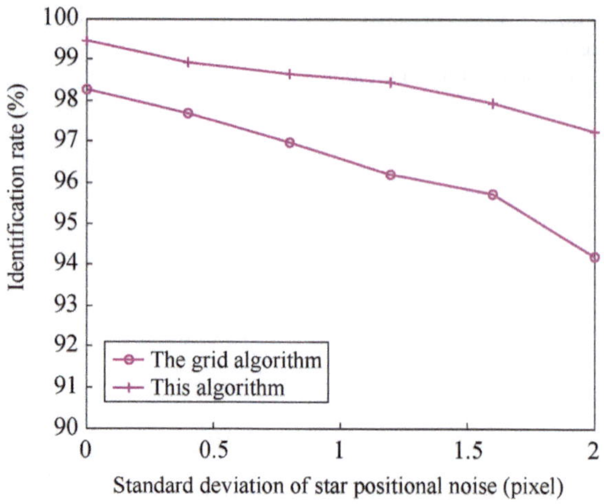

Fig. 4.12 Effect of star magnitude noise on identification

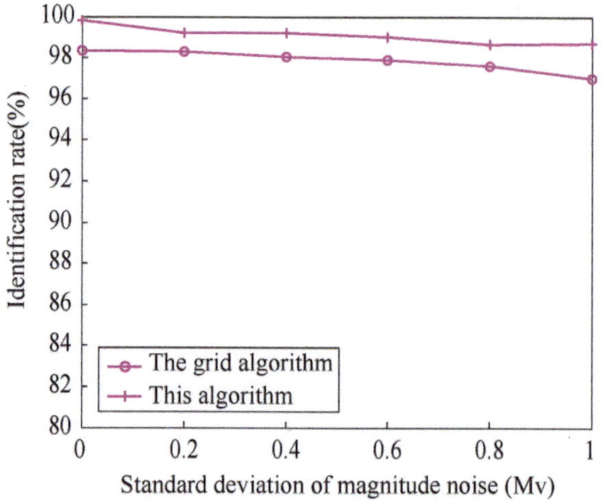

simulation. Figure 4.12 shows that two algorithms are used for identification and the statistics after different star magnitude noises are added. Each statistics is obtained from 1000 times of identification generated randomly across the sphere. The two algorithms' identification rate is barely affected by the increasing of star magnitude noise. Because magnitude information is not used in feature extraction, both algorithms demonstrate robustness to star magnitude noise.

(4) Effect of Interfering Star on Identification

With the same method in Sect. 3.2.5, a simulation experiment on the effect of an interfering star is carried out. A certain number (1–2) of "artificial stars", whose equivalent magnitude varies from 3 to 6 Mv and standard deviation of star

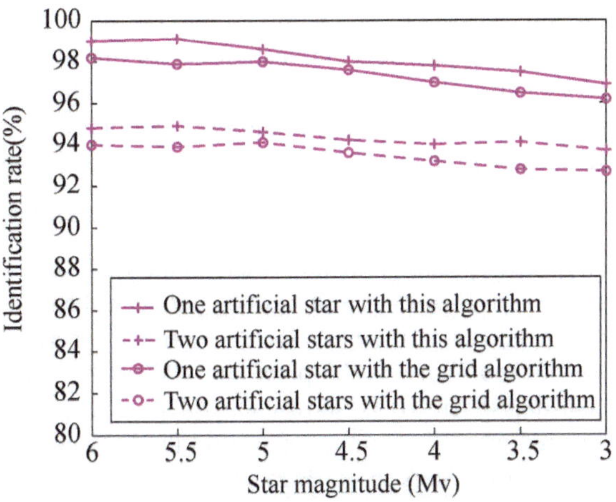

Fig. 4.13 Effect of "artificial star" on identification rate

magnitude noise is 0.2 Mv, are randomly added to the measured star image. The identification statistics are shown in Fig. 4.13. It can be seen that the grid algorithm and this algorithm are not very sensitive to artificial star magnitude. The identification is barely affected by the increasing of artificial star' brightness. The identification rate drops by about 4% with two artificial stars than that with one artificial star. Furthermore, this algorithm performs slightly better in resisting the influence of the artificial star than the grid algorithm.

In the same way, a certain number (1–2) of measured stars in the image are randomly deleted to investigate the influence of "missing stars" on identification rate. The magnitude of deleted measured stars varies from 3 to 6 Mv. Figure 4.14 shows the statistical result. It is shown that missing star magnitude, like the artificial star, has little effect on identification. Compared with that in normal conditions (no missing star, no noise disturbance), the identification rate drops by 2% with one

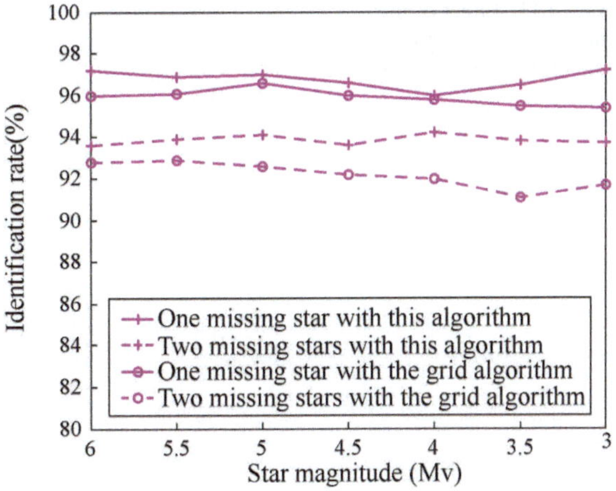

Fig. 4.14 Effect of "missing star" on identification rate

missing star and by 6% with two missing stars. Moreover, with the interference of the missing star, the identification rate of this algorithm is slightly higher than that of the grid algorithm. The main reason why the missing star results in lower identification rate is that the number of neighboring stars in the neighborhood is too small while the number of redundant matches is too large.

(5) Effect of the Number of Measured Stars in the FOV on Identification Rate

Generally speaking, the more stars in the FOV, the more likely that identification is successful. Figure 4.15 shows how the identification rate changes as the number of measured star in the FOV increases in the experiment when standard deviation of star location noise is 0.5 pixel and 1000 star images are randomly selected in the celestial sphere. It is shown in Fig. 4.15 that an identification rate of 100% is ensured when the number of stars in the FOV is over 10, while the rate drops with less than ten stars in the FOV. It is almost impossible to obtain a correct identification when the number of stars in the FOV is lower than 5. In fact, through statistics, the probability that there are over ten stars on average in the FOV is 96.62%.

The average number of measured stars in the FOV is an important parameter in star identification carried out by using "star patterns". This kind of algorithm is outstanding in performance when there are enough stars in the FOV. When stars in the FOV are sparse, the identification is difficult because not enough information can be provided to exclude redundant match(es).

(6) Identification Time and Storage Capacity

When 1000 times of identification are randomly conducted on Pentium 800 MHz PC with this algorithm, the average identification time is 11.2 ms. In the same situation, the average identification time of the grid algorithm is 10.5 ms.

In this algorithm, there are 2 bytes in the index number of each guide star in the LT. It follows that the LT needs a storage space of about 192 KB altogether. So,

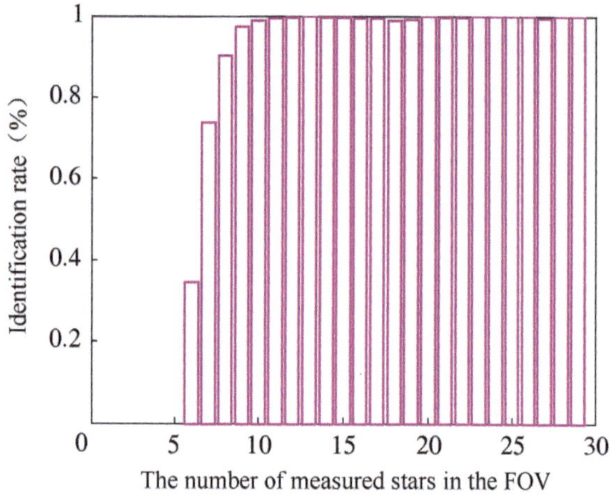

Fig. 4.15 Effect of the number of measured stars in the FOV on identification rate

plus the 3.36 bytes that the cyclic pattern needs, the star identification algorithm by using radial and cyclic star patterns requires a storage space of 196 KB altogether. In comparison, there are 73,723 records in the grid algorithm, which requires a storage space of about 144 KB.

The grid algorithm is slightly better than the present algorithm in identification time and storage space. The main reason is that star identification by using radial and cyclic star patterns is more complicated in feature extraction, storing and identification.

4.3 Star Identification Utilizing the Log-Polar Transformation Method

The differences between star identification and general image recognition are mainly twofold:

① The only feature information that can be used in star identification is star spot's position coordinates and brightness information which is not quite accurate. In some sense, star identification can be regarded as the identification of two-dimensional discrete points.

② Star identification is not completely the identification of two-dimensional discrete points. It also relies on star sensor's imaging model and parameters of the imaging system. Despite such differences between star identification and general image recognition, there are some methods in image recognition that can be referred to and applied in star identification. To realize, rotation-invariant feature extraction is the priority task in star identification by using "star patterns". Lots of research in image recognition has been conducted with and some established methods put forward.

Log-Polar transformation (LPT) is a common method used for rotation-invariant feature extraction in image recognition. Zhang et al. [3, 6, 7] introduced this method into star identification. Through transformation, a feature pattern expressed in coded strings is generated for each star. Finally, approximate string matches are employed to identify feature patterns. This section introduces the principles and implementation of the star identification algorithm by using LPT in detail and evaluates its performance through simulations.

4.3.1 Principles of Log-Polar Transformation

Schwartz [8] proposes that, between mankind's retina and visual cortex, there exists Log-Polar mapping which plays an important role in the identification of a target that is scale, shift and rotation invariant. LPT is a kind of transformation from

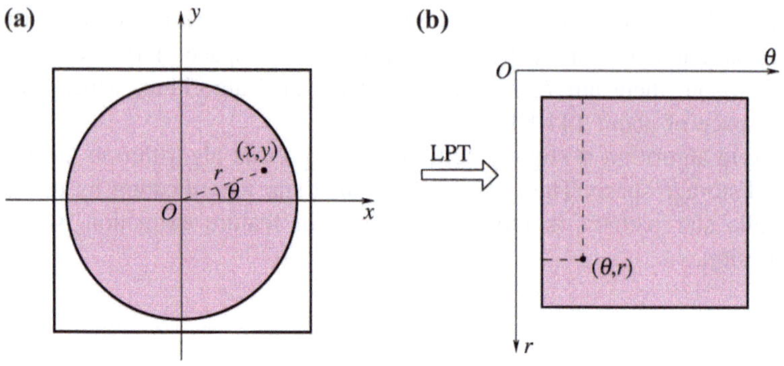

Fig. 4.16 Log-polar transformation. **a** Binary image, **b** Image after log-polar transformation

Cartesian coordinates to polar coordinates. Through mapping, the scale, shift and rotation of the target turn into one dimensional changes, greatly simplifying the problem. LPT is widely used in many areas such as moving target identification and character identification [9–11].

Denote the binary image in the Cartesian coordinate system and its LPT result image in the polar coordinate system as $f(x, y)$ and $f'(r, \theta)$, respectively. LPT (Fig. 4.16) can be defined as

$$
\begin{aligned}
r &= \ln\sqrt{x^2 + y^2} \\
\theta &= \begin{cases} \tan^{-1}(y/x) & \text{if } \ x > 0,\ \ y > 0 \\ \pi + \tan^{-1}(y/x) & \text{if } \ x < 0 \\ 2\pi + \tan^{-1}(y/x) & \text{if } \ x > 0,\ \ y < 0 \end{cases}
\end{aligned}
\tag{4.4}
$$

Through LPT, the rotation transformation in the original image turns into circular shift in θ coordinate and the scale transformation in the original image turns into shift in $r$coordinate in the polar system. Therefore, LPT is usually used to extract rotation and scale invariant features in image matching. Both images, before and after rotating, are transformed by LPT as is shown in Fig. 4.17.

4.3.2 Star Pattern Generation Utilizing the Log-Polar Transformation Method

A measured star image can be viewed as the rotation of a star image in a certain part of the celestial sphere, and star identification, in some sense, is equivalent to image matching and identification by using rotation-invariant features. If the rotated measured star image coincides with the image of a certain part of the celestial sphere, then they are considered matched. So the LPT can be employed to extract

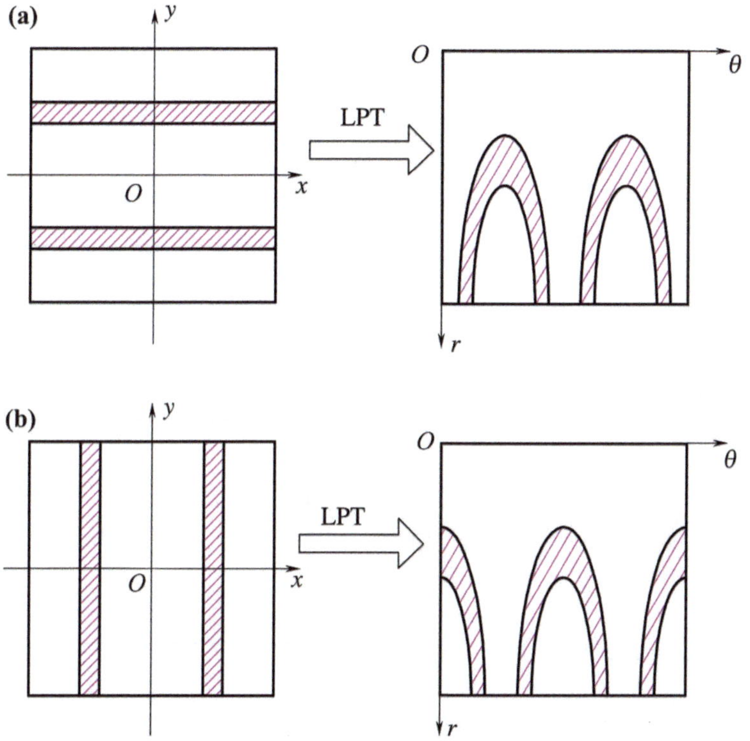

Fig. 4.17 Images' LPT results before and after rotation. **a** The original image and its LPT results. **b** The image after rotation and its LPT results

the rotation-invariant features in the star image. Different from the LPT in general image matching, the LPT in star identification only needs to transform discrete points (star spots) instead of a whole image in general cases. Additionally, in star identification, centroid position coordinates of the star spot to be identified are chosen as the coordinate origin for LPT, while in general image matching, centroids of the target to be identified are chosen as the coordinate origin.

Figure 4.18 is the illustration of star image's LPT. Figure 4.18a is a star image composed of stars in the neighborhood of the guide star s in the GSC and the image's LPT result (*s* is the origin of coordinates). Figure 4.18b is the LPT result of a measured star image, with the measured star *t* as the origin of coordinates. If the measured star corresponds to the guide star, the measured star image should coincide, after rotating by a certain angle around star *t*, with the star image composed of guide stars in Fig. 4.18a. The shift in θ coordinate in the image after LPT is circular.

With the result obtained after LPT centering around the guide star (or measured star) as the feature pattern of the guide star (or the measured star), taking the guide stars as an example, the transforming process can be illustrated as follows:

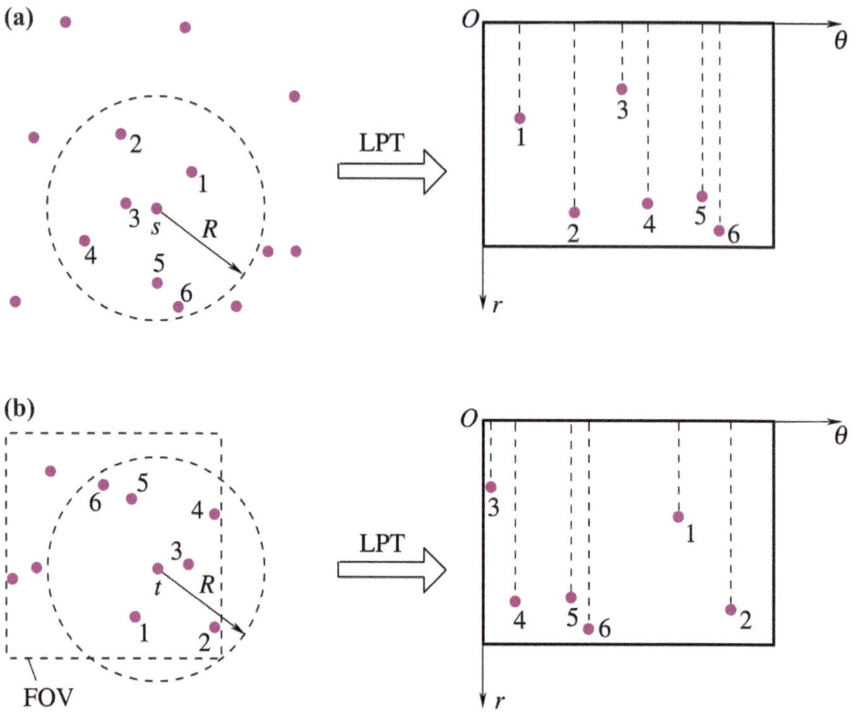

Fig. 4.18 Star image's LPT. **a** The star image of a certain celestial area in GSC and its LPT results. **b** A measured star image and its LPT results

① Take the direction vector of guide star s as the direction of star sensor's boresight, and project guide stars in the neighborhood of s with radius R (called neighboring stars of s, such as stars of number 1–6 in Fig. 4.18a), to the imaging plane (c.f. "Star Image Simulation" in Sect. 2.3). The star image obtained in this way is the original image. Apparently, guide star s is projected to the origin of the original image.

② Conduct LPT of the neighboring stars of s according to Eq. (4.4). Using a similar method, the measured star in the measured star image is transformed by LPT. Denoting the scale of the original image as $M \times N$, and that of the image after LPT as $m \times n$, then the resolution in θ and r directions are 360 °/m and R/n, respectively. Because the number of stars is much lower than $m * n$, the binary image after LPT can be expressed as an $m \times n$ sparse matrix A.

$$A(i,j) = \begin{cases} 1 & \text{at least one star at } (i,j) \\ 0 & \text{no star at } (i,j) \end{cases}$$
$$i = 1, \ldots, m, \quad j = 1, \ldots, n$$

Projecting the resulting image towards θ axis, a $1 \times m$ vector $lpt(s) = (a_1, a_2, \ldots a_1, \ldots a_m)$ is obtained. $a_i(i = 1, \ldots, m)$ is defined as follows:

① If for each $j \in (1, \ldots, n)$, $A(i,j) = 0$, then $a_i = 0$.
② If $j \in (1, \ldots, n)$ that makes $A(i,j) = 1$ exists, then a_i is equal to the minimal j that makes $A(i,j) = 1$

The $(lpt(s)$ obtained, according to the definition above, is the feature pattern of star through LPT.

If the feature pattern of star t in the measured star image is $lpt(t)$, and the feature pattern of guide star s in the GSC is $(lpt(s)$, the similarity between $lpt(t)$ and $(lpt(s)$ is defined as

$$\mathrm{sim}(lpt(s), lpt(t)) = \max_{v=1}^{m} \mathrm{same}(cs(lpt(s), v), lpt(t)) \tag{4.5}$$

Here, $cs(lpt(s), v)$ means circular shifting (in left or right directions) of $(lpt(s)$ for v bits, and same is defined as the number of matched nonzero bits in the two vectors. The bigger the same value, the more the matched nonzero bits, and the more similar these two vectors. Bigger value of $\mathrm{sim}(lpt(s), lpt(t))$ indicates higher similarity between $lpt(t)$ and $(lpt(s)$. For example, for two feature vectors with $m = 20$

$$lpt(s) = (0\ 23\ 0\ 0\ 54\ 0\ 10\ 0\ 0\ 0\ 21\ 0\ 0\ 0\ 0\ 0\ 0\ 19\ 0\ 0)$$
$$lpt(t) = (10\ 0\ 46\ 0\ 21\ 0\ 0\ 12\ 0\ 0\ 0\ 19\ 0\ 0\ 0\ 20\ 0\ 0\ 54\ 0)$$

The similarity between these two vectors is

$$\mathrm{sim}(lpt(s), lpt(t)) = \mathrm{same}(cs(lpt(s), 6), lpt(t)) = 4.$$

If measured star s matches guide star t, then

$$\mathrm{sim}(lpt(s), lpt(t)) > \xi \tag{4.6}$$

Here, ξ is a similarity threshold which is related to the number of nonzero bits in the feature vector (the number of neighboring stars constituting the feature vector).

4.3.3 Star Pattern String Coding and Recognition

(1) String Coding

The star's feature pattern $lpt(s)$ is a $1 \times m$ vector, in which most bits are zero except for a few nonzero bits. Extra capacity in the pattern vector base and more time of pattern search matching are both needed if these vectors are to be directly used for recognition. Through recoding pattern vectors obtained after LPT and denoting stars' feature pattern by strings, the storage space is saved and the search is faster. The principles of coding are as follows:

① Circularly shift, to the left direction, all zero bits before the first nonzero bit in $lpt(s)$ to the tail of the vector and thus the nonzero bit becomes the first bit of $1 \times m$ vector.

② Recode to obtain $str(s)$. Odd number bits are nonzero bits in $lpt(s)$, and even number bits are the numbers of zero between two adjacent nonzero bits in $lpt(s)$. Each character in the coded string is expressed by one byte.

Below, Lpt(s) refers to pattern vector (m = 100) after the $lpt(s)$ of guide star number 5 in the GSC. $str(s)$ is the coded string.

$lpt(s) =$ 0 0 0 0 0 0 0 0 0 0 0 0 0 0 35 0 0 0 0 0 0 53 0 0 0 0 0 0 0 0 0 44 0 0
0 0 0 52 51 0 0 0 0 0 0 0 54 48 0 0 0 0 0 0 0 0 0 0 3 0 49 0 0 0 0 0 0
0 0 0 0 0 0 0 0 53 0 0 0 0 0 0 0 0 0 0 0 0 0 0 0 0 0 29 0 0 0 0 0 0 0

$str(s) =$ 35 6 53 9 44 5 52 0 51 7 54 0 48 10 31 1 49 14 53 16 29 21

Given that rotation is transformed into circular shift through LPT, to facilitate pattern matching, the length of the coded string of guide star is extended to twice of its original length through the circular shift, while the coded string of measured star remains unchanged. The string above is extended to

$str'(s)$

36 6 53 9 44 5 52 0 51 7 54 0 48 10 31 1 49 14 53 16 29 21 35 6 53 9 44 5 52 0 51 7 54 0 48 10 31 1 49 14 53 16 29 21

(2) String Recognition

After LPT and string coding, the pattern of a star can be regarded as a word, and the set including the pattern strings of all guide stars in the whole GSC can be regarded as a dictionary. Star identification aims to find out the most similar word in the dictionary.

Denoting the pattern string as $\mathrm{str}_{1\times m}$, looking up a word in the dictionary is, in essence, to find a pattern p_i in the dictionary dict $= \{p_1, p_2, \ldots, p_N\}$ to match p_i and $\mathrm{str}_{1\times m}$. All the words in the dictionary form a text tex. Finding a word in the dictionary can be described as string matching which is to find the position of pattern string $\mathrm{str}_{1\times m}$ in the text $\mathrm{tex}_{1\times k}$. Here, $\mathrm{str}(i) \in \Re, i = 1, 2, \ldots m$, $tex(i) \in \Re$, $i = 1, 2, \ldots, k$.$\Re$ means character set and generally $k \gg m > 0$.

When the method of exhaustion in order to conduct string matching is adopted, time complexity $m(k - m - 1)$ means $m(k - m - 1)$ times of comparison are needed in the worst case situation (i.e. the whole text is to be searched). The KMP (Knuth-Morris-Pratt) algorithm [12] is a high-speed algorithm commonly used in string matching. Different from the method of exhaustion, when the matching of a

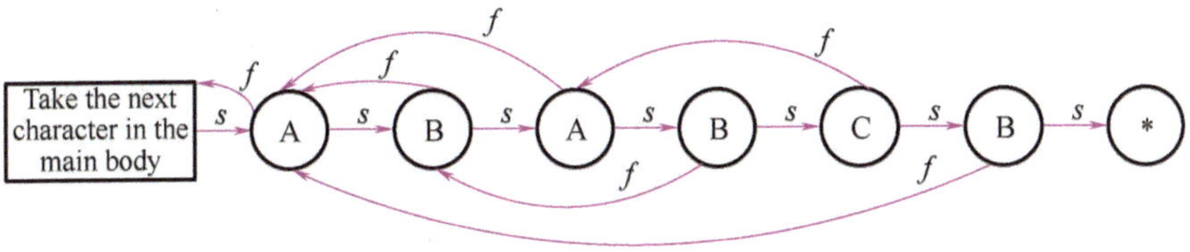

Fig. 4.19 The flow chart of KMP

certain character fails, the KMP algorithm does not simply return from the text, but makes full use of the preceding comparison information. In the KMP algorithm, a KMP flow chart, which is used to scan the text, is constructed for each pattern $str_{1\times m}$. Every node in a KMP flow chart includes only two arrows. One is called the success link, which should be followed when an anticipated character is read out in the text, while the other is called the failure link. The key to the KMP algorithm is to construct the failure link. Figure 4.19 is the KMP flow chart of pattern string $str_{1\times m}$ = "ABABCB".

The KMP algorithm reduces the complexity of string match significantly. Only $m+k$ times of comparison are needed in the worst case.

The string match algorithm used in star identification is different from general string match algorithms, so the KMP algorithm cannot be directly used for identification. The differences are mainly in the following aspects:

① The character sets these two algorithms use are different, so are the meanings of characters. The meaning of every bit in general string match is equivalent, while in star identification, meanings of bits of odd number and even number in the star pattern's string are different. Bits of odd number stand for neighboring stars' coordinate values in the r axis after LPT, and bits of even number stand for intervals in the θ axis. For star identification, string match is actually the match of strings in odd number bits. In Fig. 4.20, assuming that strings of measured star and guide star match start from (a_1, b_2), matching of a_{2i+1} and b_{2i+1} must satisfy the following simultaneously:

$$a_{2i+1} = b_{2i+1}$$

and

$$a_2 + b_4 + \cdots + a_{2i} = b_2 + b_4 + \cdots + b_{2i} \tag{4.7}$$

Apparently, this definition of character matching in string is different from that in a general case.

② In fact, string match in star identification is a kind of approximate string match. Here, "approximate" has twofold implications: firstly, measured the star's pattern is not complete. This is particularly true when the star's pattern is only a

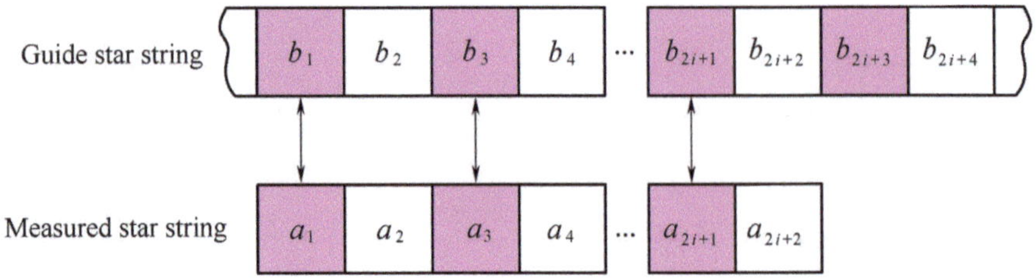

Fig. 4.20 String match in star identification

quarter, at most a half, of the capacity of its corresponding guide star's pattern when it is on the edge of the FOV. Plus star spot position error and the effect of interfering stars, measured star's pattern strings cannot completely correspond to its matched guide star's pattern strings. Secondly, due to star spot position error, principles of accurate string match cannot be applied to define character's match. Therefore, to enhance the robustness of string match, Eq. (4.7) is redefined as:

$$|a_{2i+1} - b_{2i-1}| \le 1$$

and

$$|(a_2 + b_4 + \cdots a_{2i}) - (b_2 + b_4 + \cdots b_{2i})| \le 1 \tag{4.8}$$

It is obvious that Eq. (4.8) is much looser than Eq. (4.7) in defining "match".

Approximate string match is a kind of fault-tolerant match which is widely used in many areas such as intelligent information retrieval, DNA fragments analysis, and so on. Now quite a few algorithms about approximate string match are available, for example, the agrep algorithm [13] proposed by Wu. This algorithm memorizes already matched strings with a kind of flexible bit coding. The speed is very high (just a few seconds for searching texts of several megabytes on a Sun workstation).

Du [14] has conducted detailed study on how to conduct approximate string recognition.

Based on a KMP algorithm, an algorithm of approximate string recognition suitable for the recognition of star pattern string is introduced. It follows the character matching principle defined in Eq. (4.8). General approximate string match algorithms must deal with operations like character deleting, inserting and substituting. But in star identification, a comparatively easier method is adopted. In the strings of the guide star pattern, assuming the number of matched characters (odd numbers) and mismatched characters of strings with the measured star patterns are n_{match} and n_{dismatch}, respectively. These two values are updated constantly in the process of matching. The process of identification and search is controlled by keeping track of these two values. The string match follows the principles below:

① Correct match: if n_{match} exceeds a certain threshold ξ, a correct match is found [see Eq. (4.6)], and the algorithm returns successfully. ξ is related to the number (n_{neighbor}) of neighboring stars in the neighborhood that constitute the measured star's pattern. Here, $\xi = n_{\text{neighbor}} - 2$, meaning two character' misrecognitions are allowed.

② Wrong match: if the number of mismatched characters is larger than 2, i.e., $n_{\text{dismatch}} > 2$, no correct match can be found at this position and the algorithm returns unsuccessfully. The return status is determined by the failure link constructed by the KMP algorithm. By this means, the recognition process becomes faster by quickly skipping wrong matches.

③ Searching scope: the number of neighboring stars in the neighborhood of guide star to be identified should be within a certain range. And in theory, it should be slightly larger than the number of neighboring stars in the neighborhood of measured star (for measured star on the edge of the FOV, the neighborhood scope should be a quarter to a half of its corresponding guide star's neighborhood scope). Denoting the number of neighboring stars constituting guide star's pattern string as m_{neighbor}, if $m_{\text{neighbor}} < n_{\text{neighbor}} - 2$ or $m_{\text{neighbor}} > 2n_{\text{neighbor}}$, then this guide star is skipped and the next guide star is chosen for matching. In this way, guide stars to be searched can be limited within a comparatively small scope and thus recognition becomes faster.

(3) Selection and Identification of Measured Stars

The larger the number n_{neighbor} of neighboring stars is in the neighborhood, the higher probability of correct identification it will obtain. Figure 4.21 shows the identification result of two randomly-generated star images. The bigger ○ in Fig. 4.21 stands for correctly-identified measured stars and smaller ○ for misidentified stars. It can be seen that the identification rate of the measured star near the center of the FOV which has more neighboring stars in its neighborhood is

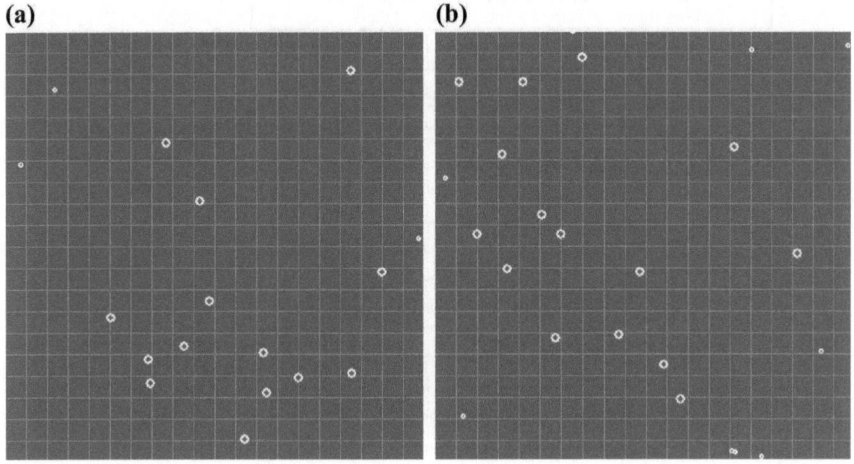

Fig. 4.21 Identification result of two randomly-generated star images

far higher than the identification rate of the measured star on the edge of FOV. Therefore, a measured star with bigger n_{neighbor} is the priority choice.

Meanwhile, a brighter star is easier to capture and is more reliable. Thus, defining each measured star as $Q = M/n_{\text{neighbor}}$ (M for star magnitude), these measured stars are ordered according to the value of Q, and the star with the smallest Q is the priority in selection for matching.

Using the method similar to the star identification algorithm above, verification is introduced into the identification. If two measured stars obtain correct identification (not the final correct identification, but the "correct identification" in the string matching described above), these two stars and their matched guide stars can be used to verify the validity of the identification. If it's verified, the identification succeeds. Otherwise, other measured stars of a lower order are selected successively for identification.

Figure 4.22 is the flow chart of the star identification algorithm found by using LPT.

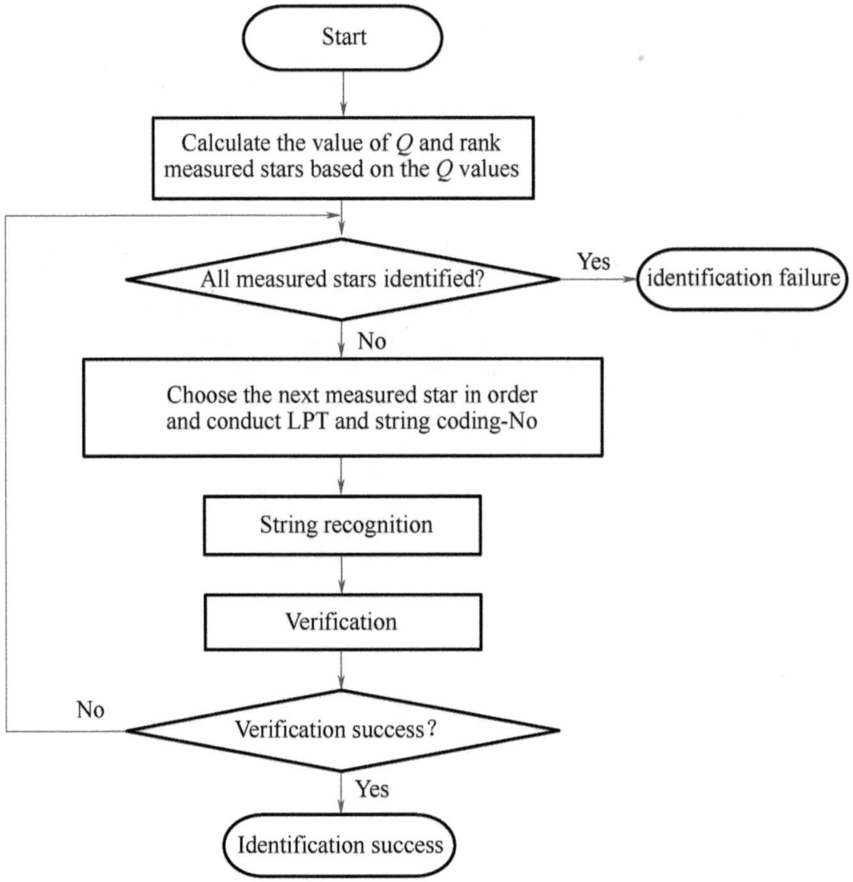

Fig. 4.22 Flow chart of the star identification algorithm by using LPT

4.3.4 Simulations and Results Analysis

In simulation, star sensor's imaging parameters are the same as the parameters used in the simulation in Sect. 3.2.5. The simulations mainly include the selection of radius R, and effects of star spot position noise, star magnitude noise and interfering stars on identification.

(1) Selection of Identification Parameters

It is a problem that the algorithm by using star patterns must solve to determine the range of neighborhood used to construct the feature pattern. If the selected pattern radius R is too small and the information is not complete, then the unique feature for each star cannot be constructed. If R is too big, there will be relatively great differences between the pattern of the measured star and the pattern of its corresponding measured star. Especially for the star close to the edge of the FOV, the pattern obtained through LPT may be just a small part of its corresponding guide star's pattern.

Select different values of R and conduct identification experiment respectively. In the experiment, values of R vary from 3° to 10°, LPT transformation parameters m and n in identification are 100 and 60, respectively, and no noise is added. Figure 4.23 represents the statistical identification result of 1000 star images randomly selected in the celestial sphere. According to the statistical result, when R is comparatively small, the identification rate is very low. The identification rate goes up with the increase of R. But after $R > 6°$, the identification rate declines. Therefore, 6° is considered a reasonable value for the pattern radius R.

(2) Effect of Star Spot Position Noise on Identification

To investigate algorithm's robustness to star location error, a gauss noise with mean = 0 and std. dev $\sigma = 0 - 2$ pixels is added to the true star position in star

Fig. 4.23 Effect of pattern radius R on identification rate

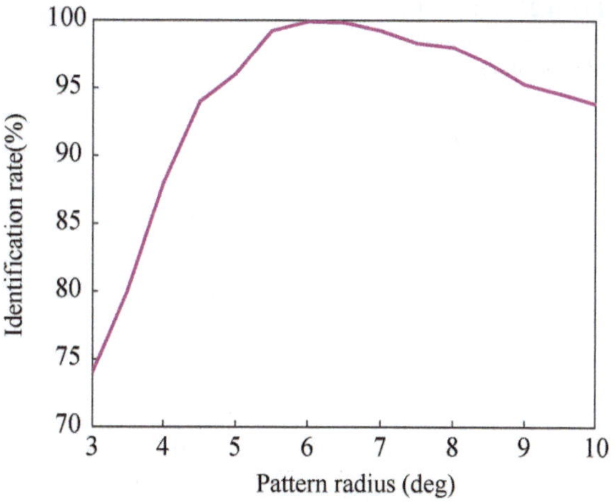

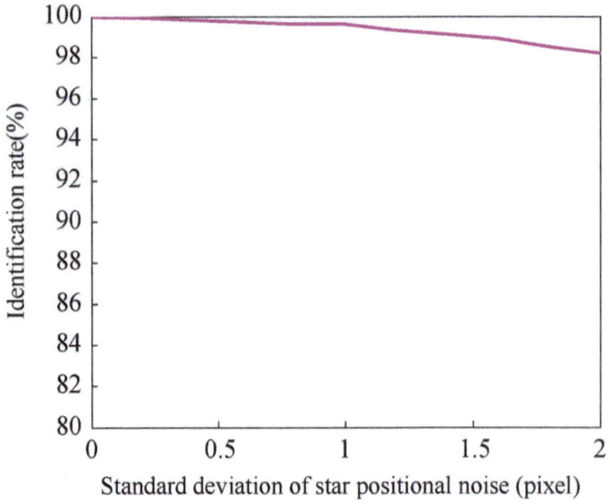

Fig. 4.24 Effect of star spot position noise on identification

image generated through simulation. Figure 4.24 shows the statistics of identification results of 1000 star images randomly selected from the celestial sphere. According to the statistical result, this algorithm demonstrates robustness to position noise, and the identification rate can still be 98% or higher when $\sigma = 2$.

(3) Effect of Star Magnitude Noise on Identification

A gauss noise with mean = 0 and std. dev = 0–1 Mv is added to the star magnitude in star image simulation. Figure 4.25 shows the statistics of identification after different star magnitude noises are added. Each of the statistics is obtained from 1000 times of random identification across the sphere. It can be seen from the statistics that star magnitude noise has a small impact on the identification rate. And the identification rate can still reach around 99% when the standard deviation of noise is 1 Mv. Thus the effect of brightness error on identification rate is negligible.

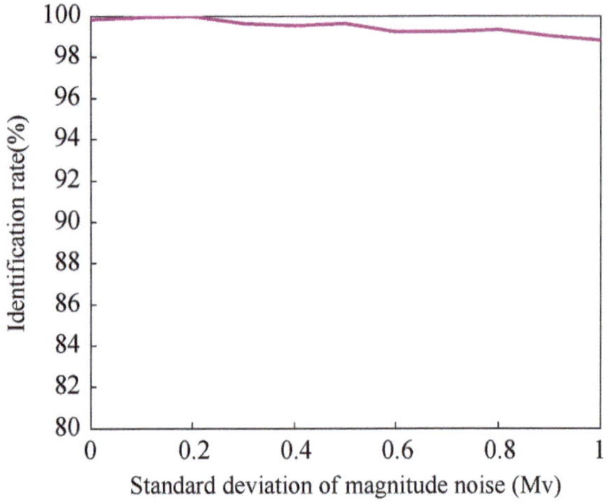

Fig. 4.25 Effect of star magnitude noise on identification

(4) Effect of Interfering Stars on Identification

A certain number (1–2) of "artificial stars" are randomly added to the measured star image. These stars' equivalent magnitudes vary from 3 to 6 Mv and the standard deviation of star magnitude noise is 0.2 Mv. The statistics of identification results are shown in Fig. 4.26. The effect of interfering stars on identification is far weaker than that in the modified triangle algorithm as is shown in Fig. 3.15. This algorithm is not sensitive, especially, to artificial stars' brightness. The identification rate still stays high even if the artificial star is relatively bright. The identification rate is lower with two artificial stars than that with one artificial star, mainly because at most two character mismatches are allowed in a string match. If over two mismatched characters are allowed, the effect of artificial stars will be improved, but more identification time will be required. Increasing the number of characters that are allowed to be mismatched will also result in the higher risk of string mismatching.

A certain number (1–2) of measured stars in the image are randomly deleted to investigate the influence of "missing stars" on identification rate. The magnitude of deleted measured stars varies from 3 to 6 Mv. Figure 4.27 shows the statistical result. It is shown that a comparatively high identification rate can be ensured with one missing star. And the rate becomes slightly lower with two missing stars. The major reason for this is that, when the number of stars in the FOV is small, deleting measured stars will make the number of neighboring stars constituting the pattern feature too small to make a correct identification.

(5) Identification Time and Storage Capacity

The average time is 11.2 ms when the identification is done randomly for 1000 times on Pentium 800 MHz PC with this algorithm. It is about twice the time used by the modified triangle algorithm in Sect. 3.2. Time is mainly consumed in match

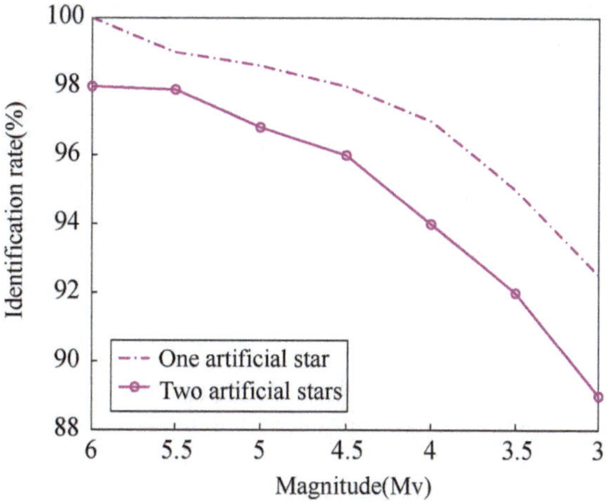

Fig. 4.26 Effect of "artificial star" on identification rate

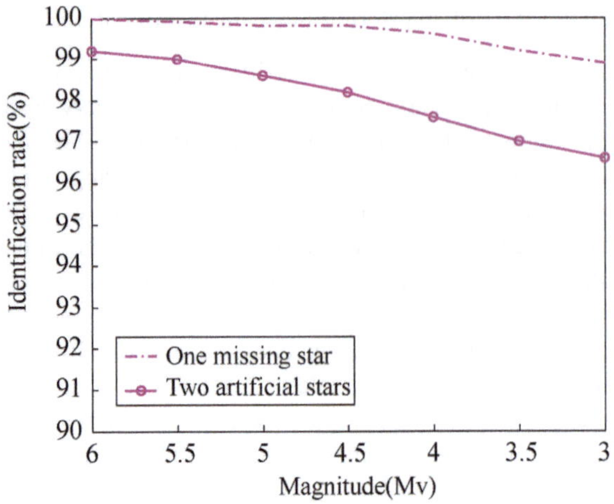

Fig. 4.27 Effect of "missing star" on identification rate

searching of strings. And the run time of this algorithm increases significantly when there are many measured stars in the FOV.

In LPT, the average length of each pattern string is about 28, so the total storage requirement of 3360 guide stars is 94 KB if one character is 1 byte. In this way, star identification algorithm, by using LPT, requires very small storage space.

4.4 Star Identification Without Calibration Parameters

Now most star identification algorithms depend on intrinsic parameters of star sensors, like focal length and principal point. In many conditions, these parameters are not accurate, or it is difficult to obtain an accurate value. Besides, the star sensor, in practical use, may be affected by shock in launch and variation of the space environment. These parameters may change or even, in serious cases, result in failure of the star identification algorithm. For example, the identification algorithm, by using angular distance, must obtain the exact values of optical system's focal length and principal points in advance. If these two parameters are not exact, the calculated angular distance will have a relatively big error calibration resulting in the algorithm's identification failure. If intrinsic parameters are not exact when the feature vector is constructed with the grid algorithm, some star spots supposed to appear in certain grid cells may enter adjacent grid cells, and the generated feature vector may change as well. Thus, most identification images can hardly obtain correct identification values when intrinsic parameters change or their exact values cannot be obtained.

To solve this problem, Zhang et al. [15, 16] proposed a star identification algorithm without calibration parameters. This algorithm introduces the scaled distance and the angle as features. These two features are only related to distances between star spots in the image. That means, accurate intrinsic parameters are not required in identification. This section introduces the implementation of this algorithm and evaluates its performance through simulation experiments.

4.4.1 Influence of Intrinsic Parameters of a Star Sensor on Star Identification

Many star identification algorithms need exact parameters of the optical system in advance, such as focal length, positions of principal points and even optical system's distortion coefficient. These parameters can be estimated with ground calibration in the lab. However during the launch, the star sensor is inevitably shocked by external forces which result in tiny changes in the relative position of the image plane and the lens's optical axis. When the star sensor is used in orbit, it is affected by the space environment variation. The optical system's parameters are thus changed. So the original parameters obtained from calibration is no longer accurate and often causes identification errors. Figure 4.28 shows the simulation, in which star sensors generate star images with the same attitude but different focal lengths. It can be seen that star spot positions change significantly with different parameters. + stands for the star spot's position when the focal length is 76.012 mm. * stands for the generated image's star spot position when the focal length is 80.047 mm.

With other constant parameters remaining the same, when there are principal point error e_p and focal length error e_f, errors of optical axis's angular distance and direction vector of star spot in each position in the image plane are calculated as shown in Fig. 4.29. It can be seen that the principal point error causes a relatively constant error in angular distance, the distribution of which in the whole star image

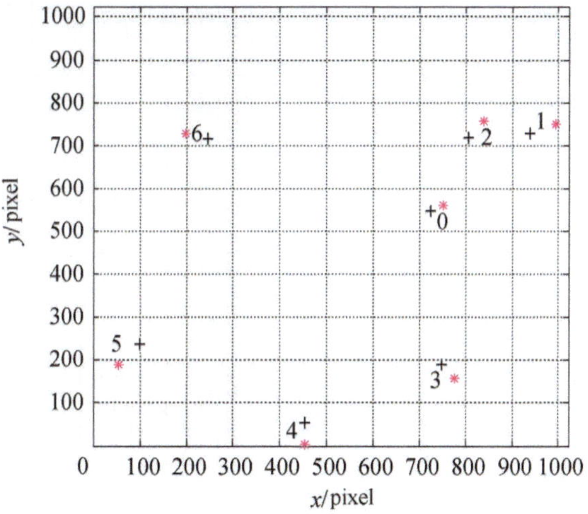

Fig. 4.28 Change of star spot position in the measured star with different focal lengths

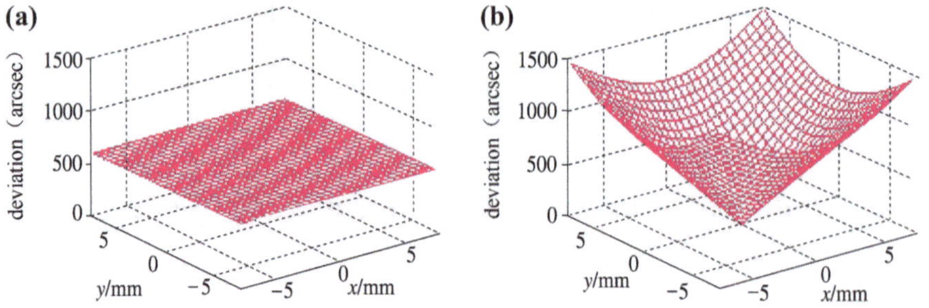

Fig. 4.29 Errors of optical axis's angular distance and direction vector of star spot due to changes of principal point and focal length

is similar. When there are errors in the focal length, the farther the star spot is from the principal point in the image plane, the bigger errors in its direction vector are. Due to these errors, wrong feature patterns may be generated and thus the star identification fails.

The position of the measured star in the grid cell needs to be judged in star identification with the grid algorithm. When feature patterns are generated, the measured star image may appear in the wrong grid cell (as is shown in Fig. 4.30) if the star sensor's focal length calibration is not exact or changes. If so, the star patterns obtained from the measured star image cannot match with the star patterns stored in the guide database correctly and thus make the identification difficult.

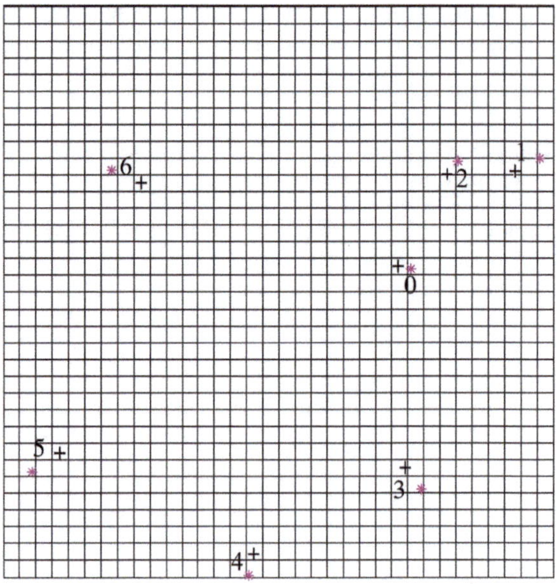

Fig. 4.30 Measured stars falling into wrong grid cells due to inaccurate focal length

4.4.2 Extraction of Feature Patterns Independent of Calibration Parameters

It can be seen, from the analysis above, that identification is affected by changes of optical parameters (the position of principal point and the focal length) used in traditional algorithms. In view of this problem, the concept of scaled distance is introduced. This distance is only related to the relative position of star spots in the image plane. The scaled distance remains unchanged no matter how these parameters change because they are not used in the algorithm. And thus the stability of star identification is ensured.

There are several situations when the star sensor captures star images. If the position of the principal point changes, the star spot's position coordinate relative to the principal point's position will shift in the image plane. If the optical system's focal length changes, the measured image will zoom in or zoom out proportionately. If the star sensor with different attitudes photographs the same celestial area, the star spot's relative position will rotate in the image plane. Therefore, it must be ensured that patterns remain unchanged in the situations above when the identification algorithm without calibration parameters is used to extract star patterns.

Star identification manifests as the matching of two-dimensional discrete points, so each star spot rather than each pixel in the whole image needs to be transformed. The process of star pattern extraction is as follows:

① Take each guide star as the primary star, search adjacent guide stars in a certain neighborhood and calculate distances between these neighboring stars and the primary star. This distance is not the angular distance defined in most identification algorithms, but the straight-line distance $R = \sqrt{x^2 + y^2}$ between two points in the image plane.

② Find out the star closest to each primary star and take it as the location star. Rotate and position the star image. To avoid the influence of binary stars and improve accuracy of rotation and positioning, the guide star at a certain distance R_0 is often selected as the location star.

③ Transform the star image. Calculate the straight-line distance R_b between the location star and the primary star in the plane and take this distance as the standard. Ratio R_{ri} of the distance R_i between one of other neighboring stars and the primary star and the standard distance R_b ($R_{ri} = R_i/R_b$) is defined as the distance between this neighboring star and the primary star and is called scaled distance. Starting from the location star, the counter-clockwise direction surrounding the primary star is in the positive direction. Angles between each neighboring star and the location star are calculated successively and denoted as the neighbor star's angle coordinate. Through the transformation above, the original star image is in the coordinate system with axis θ—R.

④ Construct the feature vector of the primary star. The coordinate system with axis θ—R is equally divided into M sectors in the direction of scaled distance, and is equally divided into N sectors in the angle direction. The transformed star image is divided into cells, and meanwhile a new $M \times N$ pattern vector pat(S) is constructed. If there are stars in a certain cell, the corresponding value of pat(S) is 1. Otherwise, it's 0. The feature vector can quantify the distribution of the primary star's neighbor stars into a vector constituted by 0 and 1.

$$\text{pat}(S) = (b_1, b_2, \ldots b_i \ldots b_{M \times N}) \tag{4.9}$$

Here, when there are stars in the cell corresponding to b_i, the value of b_i is 1, otherwise, it's 0.

Figure 4.31 shows the process of constructing feature vectors. The obvious distinction between the above-described feature construction method and the grid algorithm, lies in the fact that the grid algorithm directly uses star spot's coordinates to construct features and these features will change after image zooming in or out. By comparison, the scaled distance used to construct features in the star identification algorithm without calibration parameters is independent of imaging system parameters. During feature construction, all scaled distances between the closest

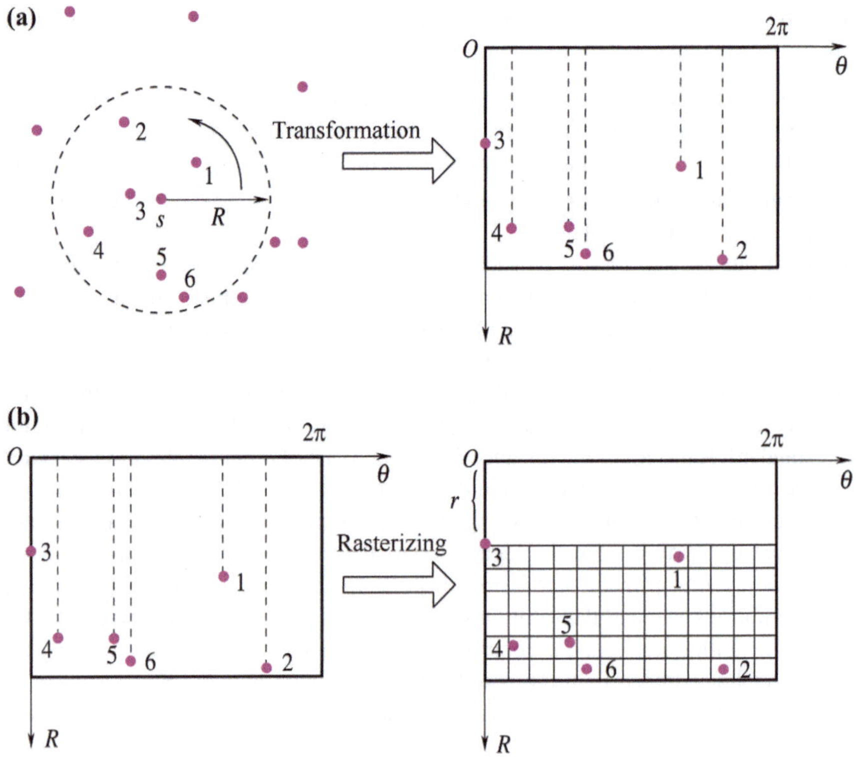

Fig. 4.31 Construction process of the star image pattern

Fig. 4.32 Star pattern's storage structure

Number	Star number
24	472 1077 1117 1214 ...
27	62 474 1154 1302 ...
21	660 1265 1296 1340 ...

neighboring stars and the primary star are 1. Because the scaled distance is a relative value, it will not change with the image zooming in or out. So it is an invariant. The angle is determined by the location star, so it's not related to the attitude when images are captured and is a rotation invariant. In addition, the positional relationship between star spots rather than unreliable information like brightness is used when feature vectors are constructed. Therefore, vectors constructed with the above method enjoy great stability.

To identify a star, the one and only distinctive feature of the star needs to be extracted. Generally, features of stars in the guide star pattern database can be pre-computed and stored in the star sensor's memory. Then these data are directly read out during identification. When the guide star pattern database is constructed, all stars in the GSC needs to be traversed, and all of their features are calculated and stored in order.

Every star image is divided into $M \times N$ grid cells. To save storage space, a storage sequence is set for each grid cell, so there are $M \times N$ sequences in all. Each grid cell stands for a position in the guide star's neighborhood. If some guide star has neighboring stars in this position, the index number of this guide star is recorded into the grid cell's sequence and the counter of this sequence is added by 1. Figure 4.32 shows the sequence's storage format. Here, Number stands for the number of stars recorded in this sequence, and Star Index for the star index of guide stars recorded in this sequence which indicates these guide stars will, as neighboring stars, fall into the grid cell expressed by this sequence. Data in Fig. 4.32 is a part of the pattern database when $M = 30$ and $N = 80$. The pattern database's structure is simple, each entry of which only includes the number of stars recorded in the sequence and the guide stars' index numbers in ascending order. During the construction of the pattern database, the guide star's index number, which appears repeatedly in the same entry, should be eliminated.

4.4.3 Matching and Identification

The process of star identification without calibration parameters is mainly divided into two steps: firstly the initial match (approximate match) is conducted to reduce the searching range into a relatively smaller order, and then candidate stars are quickly divided into groups on the basis of FOV constraint. Finally the correct match is obtained.

(1) Initial Match

Take each guide star in the measured star image as the primary star, and calculate the radial distance R_i between this primary star and its neighbor stars based on the Eq. $r = \sqrt{x^2 + y^2}$. Also record the closest neighbor star's index number and the distance R_b between it and the primary star. There must be a certain distance between this closest neighboring star and the primary star. Calculate the scaled distances ($R_{ri} = R_i/R_b$) between other neighboring stars and the primary star. The scaled distance must be within a certain range, and neighboring stars beyond this range are not considered. The primary star's feature vector is constructed according to the neighboring stars' distribution. Most neighboring stars of the measured stars near the center of the image can appear in the image and the measured stars' patterns are comparatively complete. But for stars close to the edge of the image, their patterns may be missing. Figure 4.33 shows the pattern constructed for a certain measured star.

In the initial match, the screening range covers all guide stars in the GSC since there is no prior information. Assume that there are Ns guide stars in all, assign a counter $CT_1, CT_2 \ldots CT_{Ns}$ for each guide star and the initial values of these counters are 0. If a certain primary star has neighboring stars in the i-th cell, read out data in the i-th line of the guide pattern database and add 1 to each of the counters of all corresponding stars in the i-th line. The rest of the neighboring stars are dealt with similarly. Find out the maximum value of these counters after all the neighbor stars of this primary star are scanned. The guide star corresponding to this maximum value is the match for the primary star. Due to errors of star spot position in imaging, or pattern missing of stars close to the edge, the pattern of the measured star cannot be completely identical with that of the guide star, so the guide star

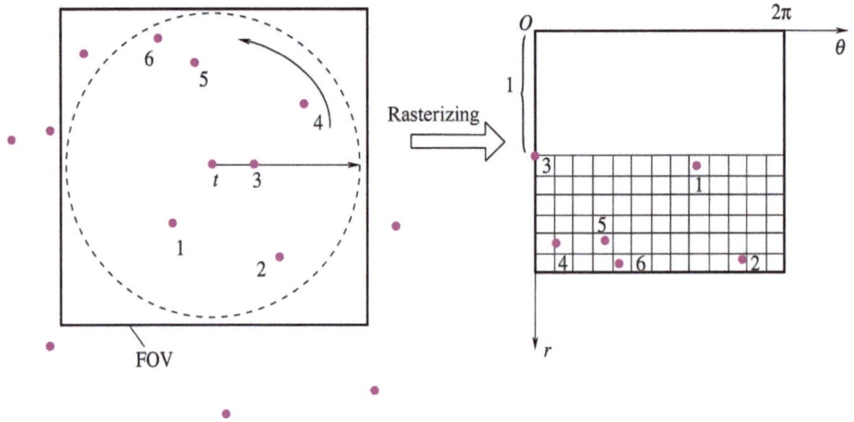

Fig. 4.33 Pattern construction of measured star image

corresponding to the maximum value is not unique, and there are often several guide stars matching the measured star. The initial match aims to narrow down the searching range and the exact match should be conducted as follows.

(2) Fast Grouping

Generally, after the result of an initial match is obtained, the angular distance between each two stars waiting for selection is calculated to judge the relationship between them and get the final unique result. With the initial screening, the grid algorithm adopts an FOV constraint to judge which stars are in the same celestial area. If the angular distances between a star waiting for selection and most other stars are beyond the FOV, this star is considered redundant and can be eliminated from the waiting list. Assume that the stars screened initially are randomly distributed in the celestial area, then the part of celestial area where most screened guide stars are located is the area to be measured by the star sensor. This algorithm needs to compute the angular distances between every two stars waiting for selection. So if there are many stars on the waiting list, a great amount of calculation and much more identification time may both be required. If there are n stars in the star image, and each of them has m_i stars waiting for selection, then the number of calculations of angular distance is $\sum_{1\le i<j\le n} m_i m_j$.

To speed up grouping, a new method for fast grouping is adopted. Assign a counter for each sub-block, and the original states of these counters are 0. Judge the sub-blocks where all stars waiting for selection are located and add 1 to the corresponding counter. Divide these stars into groups. The value of each counter stands for the number of stars waiting for selection in this sub-block. The stars waiting for selection can be considered as random dots distributed with equal probability. All these stars are distributed randomly in the celestial sphere and the part with the highest concentration of random dots is the area to be measured in the FOV. The sub-block that has the biggest counter value stands for the area where the random dots concentrate. In general, the number of stars waiting for selection is the same as that of the measured stars in this sub-block and eight adjacent sub-blocks. To ensure correct identification, the screened stars waiting for selection are examined by FOV constraints. If so, the number of calculations of the angular distance is no more than C_n^2. Assume that there are ten measured stars in the image and each star has five stars waiting for selection. That means $n = 10$ and $m_1 = m_2 = \cdots = m_n = 5$. The calculation amount required by the method of fast grouping is 4% of the original method $(C_n^2/\sum_{1\le i<j\le n} m_i m_j = C_{10}^2/(C_{10}^2 \times 5 \times 5) = 4\%)$.

Figure 4.34 is the flow chart of star identification without calibration parameters.

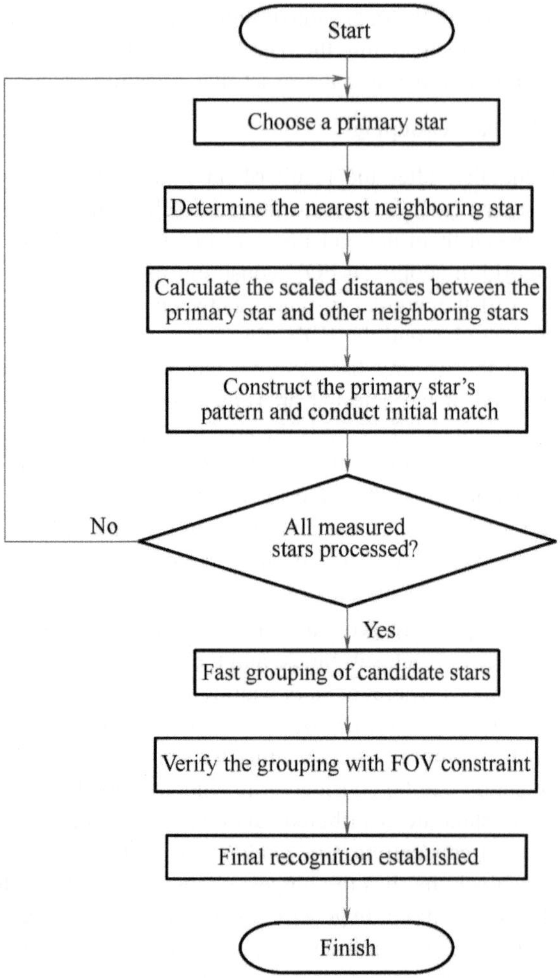

Fig. 4.34 Flow chart of star identification without calibration parameters

4.4.4 *Simulations and Results Analysis*

Imaging parameters of the star sensor in simulations are as follows: the size of the FOV is 10.8° × 10.8°, the focal length of optical system is 80.047 mm, the pixel size is 0.015 mm × 0.015 mm, and the pixel resolution is 1024 × 1024. Select stars brighter than 6 Mv from the SAO J2000 basic star catalog to construct a GSC and generate a corresponding pattern vector database. Simulations are conducted on an Intel Pentium4 2.0 GHz computer. The simulations mainly include a selection of radial scaled distance radius, and the effect of calibration parameters error, star spot position noise, star magnitude noise and the number of stars in the FOV on identification.

(1) Selection of Radial Scaled Distance Radius

Radial scaled distance radius is an important parameter. It is the largest scaled distance between the neighbor stars and the primary star during the construction of feature vectors. Radial scaled distance radius determines which neighboring stars can be used in construction of the primary star's feature vector. When the radial scaled distance radius is determined, the division grades of radial scaled distance and the cyclic angle need to be analyzed. In other words, the optimum values of M and N need to be determined, so that the highest identification rate can be achieved. The values of M and N have a great effect on this algorithm's performance. If there are too few division grades, several stars may appear in the same cell and the distribution of neighbor stars cannot be fully taken into consideration. On the contrary, too many division grades may result in very small cells. In this case, star spots in some cells may enter adjacent cells when the star spot position noise is comparatively large. If so, the division result of the measured star images does not comply with that of the guide star images and thus correct identification cannot be achieved. Therefore, only when the division grades of radial scaled distance and cyclic angle M and N are controlled within a reasonable range can correct identification be ensured.

To determine the relationship between the radial scaled distance radius and the two division grades M and N, firstly set the initial radius of radial scaled distance as 10, the initial division grade of cyclic angle N as 50, and the division grades of radial scaled distance radius as 10, 20, 30, 40, 50 and 60. Investigate the identification of 1000 star images with different grades. Figure 4.35 indicates how the identification rate of star identification without calibration parameters changes with different division grades. It can be seen that the identification rate is comparatively high if the radial scaled distance is divided into 30 grades.

According to the analysis above, the radius of radial scaled distance is still 10 and the division grade M of radial scaled distance is 30. On this basis, determine the

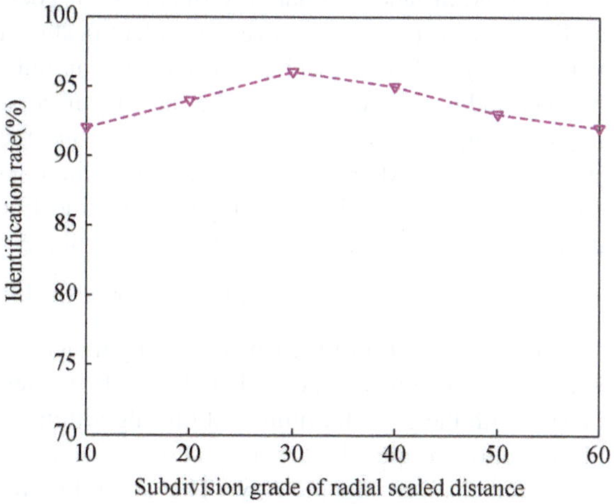

Fig. 4.35 Effect of division grade of radial scaled distance on identification rate

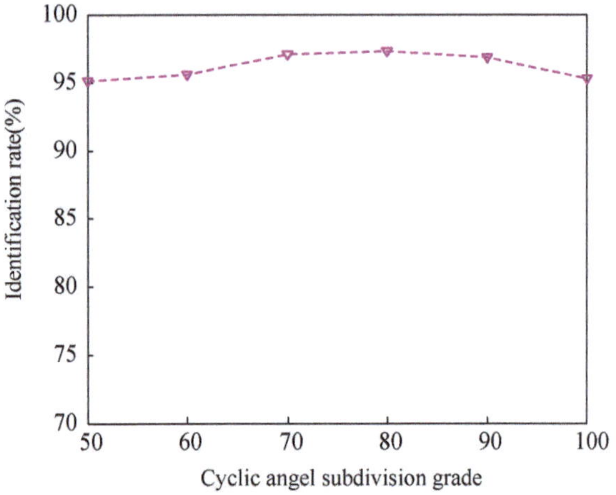

Fig. 4.36 Effect of division grade of cyclic angle on identification rate

optimum division grade of the cyclic angle. Division of the cyclic angle is actually dividing a circle into several equivalent sectors since the range of cyclic angle is 360. Suppose N = 50, 60, 70, 80, 90 and 100 and then investigate the identification. Figure 4.36 shows the identification rates with different division grades. It shows that a comparatively high identification rate can be achieved when the division grade of the cyclic angle is 80.

It can be seen, from the analysis above, that the algorithm can achieve the highest identification rate when the radial scaled distance radius is 10, the division grades of radial scaled distance and cyclic angle are 30 and 80, respectively. When the radius of radial scaled distance changes, the corresponding division grade of scaled distance should be adjusted according to this scale while the division grade of cyclic angle remains the same. Only in this way can the algorithm of this radius realize the highest identification rate.

It is shown in Fig. 4.37 that the identification rate is low when the radius of radial scaled distance is relatively small, and it goes up with the increase of radial scaled distance radius. However, the identification rate changes gently when the radius is larger than 12, which indicates that an unlimited increase of radial scaled distance radius will not lead to the increase in identification rate. Instead, it will require a larger capacity of pattern database and thus bring more pressure on the storage capacity. After every factor is taken into consideration, it is concluded that the algorithm can achieve the highest identification rate when the radial scaled distance radius is 12, $M = 14/12 \times 30 = 35$, and $N = 80$.

(2) Effect of Focal Length Calibration Error on Identification Rate

Assume that the error of the lens focal length increases from −1 to 1 mm, 1000 star images are randomly generated under each of these error grades. Identify these images with the grid algorithm and this algorithm respectively, and then record the identification results. Figure 4.38 shows that the identification rate of the grid algorithm changes significantly. The algorithm almost fails when the error is

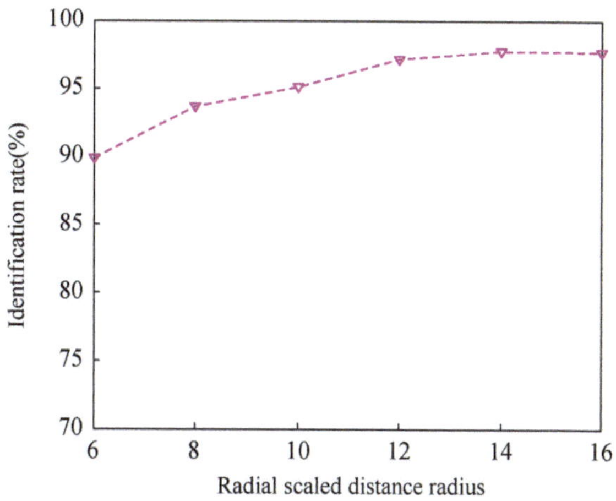

Fig. 4.37 Effect of radial scaled distance radius on identification rate

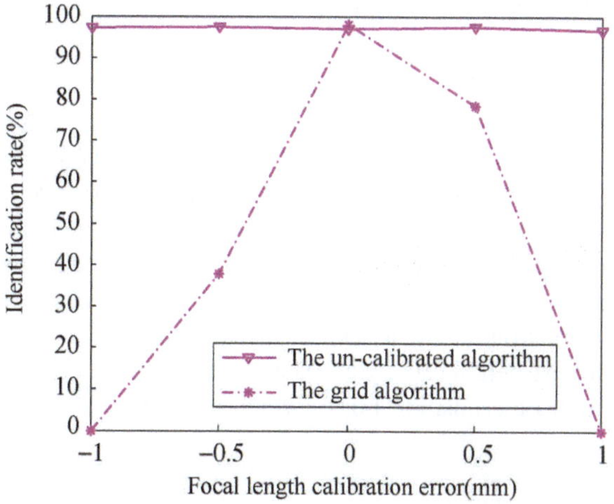

Fig. 4.38 Effect of focal length calibration error on identification rate

relatively big. By comparison, star identification without calibration parameters enjoys great stability and can still get a high identification rate, even if the focal length error is very big.

(3) Effect of Error in Principal Point Position Calibration on Identification Rate

Figure 4.39 shows the identification results of these two algorithms, when the error of the principal point's position varies from −20 to 20 pixels. It can be seen that the star identification algorithm without calibration parameters can achieve a stable identification rate when the principal point's position changes, while the identification rate of the grid algorithm drops with the increase of principal point position error.

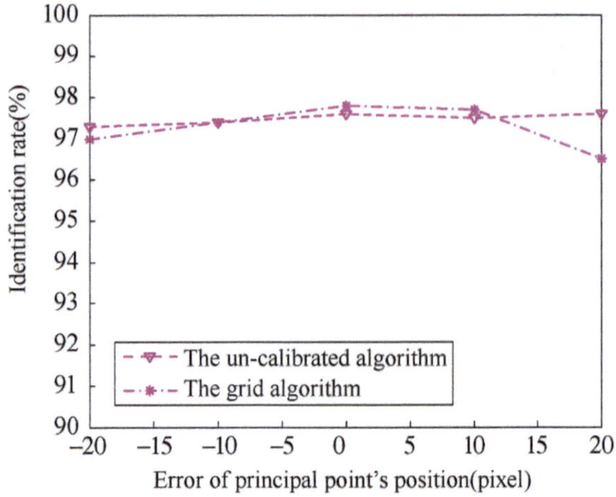

Fig. 4.39 Effect of principal point position calibration error on identification rate

(4) The effect of Star Spot Position Noise on Identification Rate

To investigate the influence of star spot positioning error on identification rate, a gauss noise with mean = 0 and std. dev $\sigma = 0 - 1$ pixel is added to the true star position in the simulated star image. Figure 4.40 shows the identification results of 1000 randomly generated star images with various kinds of noises. It can be seen that the algorithm without calibration parameters demonstrates great performance in the identification of star images when the star position noise is small. However, the identification rate drops rapidly and is lower than that of the grid algorithm when the noise is relatively big. This is because the identification algorithm without calibration parameters uses scaled distance, which is calculated on the basis of the closest neighboring star. The position of the closest neighboring star will change due to the position noise, which may result in comparatively big errors in scaled

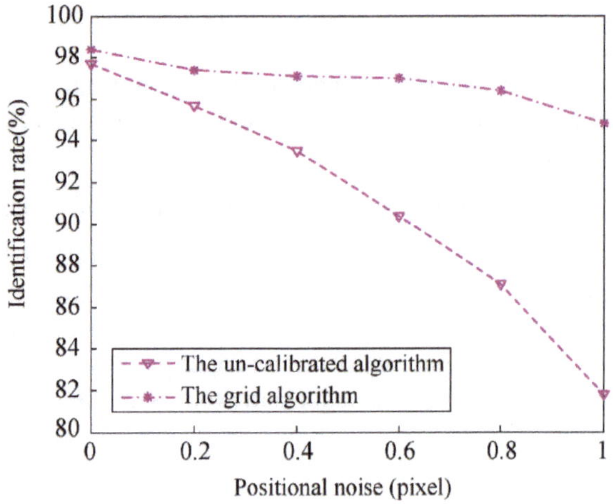

Fig. 4.40 Effect of star spot position noise on identification rate

distances of other measured stars. Therefore, the identification rate of this algorithm is lower than that of the grid algorithm. According to practical experience, error in star spot centroiding is generally smaller than 0.5 pixels, so the identification rate of the algorithm without calibration parameters is at least 92% in practical use, even if there exist errors in star spot positions.

(5) Effect of Star Magnitude Noise on Identification Rate

To investigate the performance of an identification algorithm under the influence of star magnitude noise, a noise with mean = 0 and variance = 0–1 magnitude is added to the generated star image. Two algorithms are used respectively for identification after different star magnitude noises are added. Figure 4.41 shows the statistics of identification. Each of the statistics is obtained from 1000 times of identification generated randomly across the whole celestial sphere. The two algorithms' identification rate is barely affected by the increase in star magnitude noise. Because brightness information is not used in feature extraction, both algorithms demonstrate strong resistance to star magnitude noise.

(6) Effect of the Number of Stars in the FOV on Identification Rate

Similar to the grid algorithm, star identification without calibration parameters takes distribution of stars surrounding a certain star in the image as this star's pattern. Therefore, the number of measured stars in the image should meet a certain demand. In the design of the star sensor, there should be enough measured stars in the FOV at each time of image capturing so that the identification rate of the star pattern-based algorithm can be ensured. Figure 4.42 indicates that both the grid algorithm and this algorithm do not perform well when the number of stars in the FOV is smaller than 6. An identification rate lower than 60% cannot completely meet the demand in practical use. However, the identification rates of these two algorithms can reach 95% or higher, when there are over 8 measured stars in the FOV.

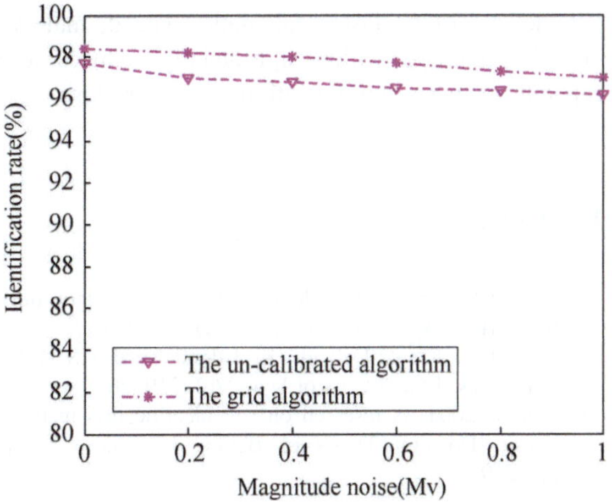

Fig. 4.41 Effect of star magnitude noise on identification rate

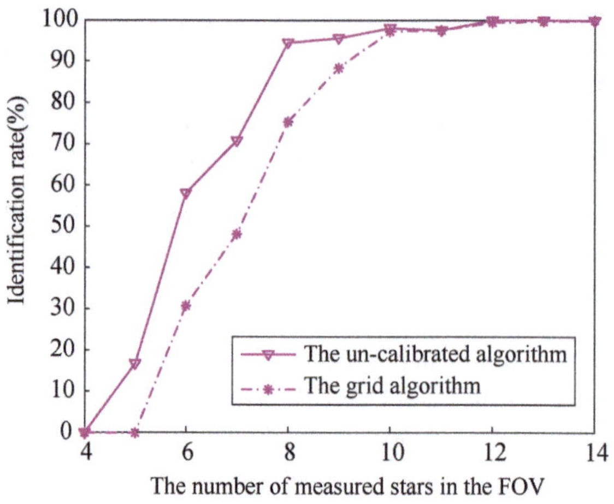

Fig. 4.42 Effect of the number of stars in the FOV on identification rate

Table 4.3 Comparison of identification time and storage capacity between the star identification algorithm without calibration parameters and the grid algorithm

Method / Performance	Star identification without calibration parameters	Grid algorithm
Average identification time (ms)	7.3	10.2
Storage capacity (KB)	372	362

(7) Identification Time and Storage Capacity

The star identification algorithm without calibration parameters and the grid algorithm are operated on the same hardware platform. The comparison of these two algorithms' identification time and storage capacity is shown in Table 4.3. It can be seen that, compared with the grid algorithm, the algorithm without calibration parameters enjoys faster calculation speed, mainly because of the use of a fast grouping method which significantly reduces identification time. The storage capacity of these two algorithms is rather close.

References

1. Padgett C, Kreutz-Delgado K (1997) A grid algorithm for autonomous star identification. IEEE Trans Aerosp Electr Syst 33(1):202–213
2. Padgett C, Kreutz-Delgado K, Udomkesmalee S (1997) Evaluation of star identification techniques. J Guid Control Dyn 22(2):259–267
3. Wei X (2004) A research on star identification methods and relevant technologies in star sensor. Doctoral Thesis of Beijing University Aeronautics and Astronautics, Beijing, pp 62–96

4. Zhang G, Wei X, Jiang J (2008) Full-sky autonomous star identification based on radial and cyclic features of star pattern. Image Vis Comput 26(7):891–897
5. Wei X, Zhang G, Jiang J (2004) A star map identification algorithm using radial and cyclic features. Opto-Electr Eng 31(8):4–7
6. Wei X, Zhang G, Jiang J (2009) A star identification algorithm based on log-polar transform. AIAA J Aerosp Comput Inf Commun 6(8):483–490
7. Wei X, Zhang G, Jiang J (2006) A star identification algorithm based on log-polar transformation. Opt Tech 32(5):678–681
8. Shwartz EL (1977) Spatial mapping in the primate sensory projection: analytic structure and relevance perception. Biol Cybern 25:181–194
9. Tao Y, Ioerger TR, Tang YY (2001) Extraction of rotation invariant signature based on fractal geometry. Int Conf Image Process 1:1090–1093
10. Kageyu S, Ohnishi N, Sugie N (1991) Augmented multi-layer perceptron for rotation-and-scale invariant hand-written numeral recognition. IEEE Int Joint Conf Neural Netw 1:54–59
11. Chen Zhaoyang, Ding M, Zhou C (1999) Target searching method based on log-polar coordinate mapping. Infrared Laser Eng 28(5):39–42
12. Knuth DE, Morris JH, Pratt VR (1977) Fast pattern matching in strings. SIAM J Comput 6 (1):323–350
13. Wu S, Manber U (1992) Agrep—a fast approximate pattern-matching tool. In: Usenix winter 1992 technical conference, San Francisco, pp 153–162
14. Du MW, Chang SC (1994) An approach to designing very fast approximate string matching algorithm. IEEE Trans Knowl Data Eng 6(4):620–632
15. Yang J (2007) A research on star identification algorithm and RISC technology application. Doctoral Thesis of Beijing University Aeronautics and Astronautics, Beijing, pp 34–47
16. Yang Jian, Zhang G, Jiang J (2008) A star identification algorithm for un-calibrated star sensor cameras. Opt Tech 34(1):26–32

Chapter 5
Star Identification Utilizing Neural Networks

Star identification based on "star pattern" is a typical problem in pattern recognition. In this case, each guide star is assigned with a unique pattern and thus has a corresponding feature vector. The task of star identification is to extract the feature patterns of the measured star and put it into the class where the guide star has similar feature patterns.

Neural networks, which have been widely used, are common methods to solve problems in pattern recognition and play an important role in the field of pattern recognition. Neural networks have a strong ability to approach nonlinear functions, or what is called the ability of nonlinear mapping. They also enjoy a series of advantages, such as parallel operation, distributed information storage, strong fault-tolerant ability, and a self-adaptive learning function. All these characteristics are the foundation which enables neural networks to be applicable for pattern recognition, especially the learning function and fault-tolerant ability, which can play a unique role in the identification of undetermined patterns. Star identification by using neural networks presents [1–3] mainly has the following characteristics: The features of star patterns are reflected as the connection strength between the weights of each neuron, which means the pattern vector database is replaced by a weight matrix. The matching between measured star patterns and guide star patterns can be finished after one single matching. There is no need for traversal matching for all star patterns.

This chapter first introduces the basic principles of neural networks and on this basis further introduces two star identification algorithms based on neural networks: the neural network star identification algorithm by using the star vector matrix feature and the neural network star identification algorithm by using the mixed feature. The working principles and implementation of these two algorithms are described in detail. Finally, an evaluation of their performances is carried out through simulation experiments.

G. Zhang, *Star Identification*, DOI 10.1007/978-3-662-53783-1_5

5.1 Introduction to Neural Networks

Neural networks are simulations of the information processing ability of human brains and have become a significant research field in artificial intelligence. Application of this in the field of star identification by star sensors has been achieved. In this section, the basic concepts, characteristics and principles are presented.

5.1.1 Basic Concepts of Neural Networks

Artificial neural networks (ANNs), also known as neural networks (NNs), are algorithmic mathematical models [4] which process information in distributed and parallel ways by simulating the characteristics of cerebral neural networks. These networks depend on the complexity level of the system, adjust the interconnection of massive inner nodes, and thus achieve the purpose of information processing. Artificial neural networks have the ability of self-learning and self-adaption. The networks can analyze and grasp the potential laws between a batch of previously provided corresponding input–output data. Based on these laws, output data can be calculated when new input data is given. This process of learning and analyzing is called "training."

ANNs are networks in which massive artificial neurons interactively connect with each other and every neuron is only a very simple information processing unit. The structure of a strong interconnection network determines that neural networks are equipped with a strong fault-tolerant ability, making it easy for the networks to "learn." By simply adjusting the connection form and strength, neural networks are able to remember new information and update the database. It is evident that human brains gain new knowledge and information by the stimulation of plenty of examples. ANNs work in the same way as human brains. Through the training of extracted samples from concrete problems, ANNs capture the inherent attributes of the problem and then apply the trained network to calculate other examples of the same problem. This is the learning process of ANNs.

In 1943, psychologist McCulloch and mathematical logician Pitts built neural networks and a mathematical model and called it the MP Model. On the basis of this MP Model, they put forward the formal mathematical description and network structure of neurons and finally proved that one single neuron also has a logic function. Thus, they began a new era of research on ANNs. In 1949, psychologist Hebb proposed that the connection strength between synapses is variable. In 1960s, ANNs were further developed with the introduction of an improved neural network model, including perceptrons and self-adaptive linear elements. After analyzing the functions and limitations of neural networks represented by perceptrons, Minsky et al. published *Perceptron* in 1969 and indicated that perceptrons failed to solve problems in higher order predicates. Their argument greatly influenced research on neural networks. Significant achievements in serial computers and artificial intelligence were also made during that time. The importance and urgency to develop new computers and new approaches in artificial intelligence were eclipsed. Therefore, research on ANNs was at low ebb. However, there were still researchers in the ANNs field who continued to devote themselves to the study and put forward the Adaptive Resonance Theory (ART Net), self-organizing maps and cognition networks, as well as research on the mathematical theory of neural networks. All the research above helped to lay the foundation for the study and development of neural networks.

In 1982, Hopfield from the California Institute of Technology introduced the Hopfield neural network model, with the concept of "energy calculating" and the judgment of network stability. In 1984, he put forward the continuous-time Hopfield neural network model, which was a breakthrough in neural computer research. This model also offered a new approach for neural networks to be applied in associative memory and calculation optimization and significantly promoted the research on neural networks. In 1986, Rumelhart, McClelland, and other people introduced the "Error Back Propagation Algorithm," known as the BP algorithm, of multilayer feedforward networks. It helped the multilayer feedforward network model to become more practical. On the basis of the multilayer perceptron network, several feedforward networks in different forms were derived in succession, for example, the radial basis function network, functional link network, and so on. Since the 1990s, as the research on neural networks moves on, hardware and soft computing tools have appeared and neural network technology has become widely applied in the fields of pattern recognition, signal processing, and robots.

5.1.2 Basic Characteristics of Neural Networks

ANNs are nonlinear and self-adaptive systems which consist of massive processing units. The idea of the artificial neural network was proposed on the basis of research findings in modern neuroscience. By simulating the methods of processing and memorizing information performed by human brains, ANNs try to realize information processing. ANNs have four basic characteristics:

① Nonlinearity. Nonlinear relationships are universal in nature. The intelligence of human brains is one of the nonlinear phenomena. Artificial neurons are kept either in an activated state or an inhibited state, and this phenomenon is shown as a nonlinear relationship in mathematics. Networks composed of threshold neurons have better performance and can improve their fault-tolerant ability and expand their storage capacity.

② Free of limitation. Usually, a neural network is connected by multiple neurons. The overall behavior of a system is not solely dependent on the characteristics of one neuron, but might be more decisively determined by the interaction and interconnection between units. Different units form massive connections to simulate the nonlimitation capability of human brains. Associative memory is a typical example of nonlimitation.

③ Flexibility. ANNs possess the ability of self-adaptation, self-organization, and self-learning. Not only can the information be processed by the neural networks change, but the nonlinear dynamic system is continuously transforming during information processing. The iterative process is often used to describe the evolutionary process of the dynamic system.

④ Non-convexity. Under certain conditions, the evolutionary direction of a system is determined by a specific state function, such as energy function, whose extreme value is at a comparatively stable state. Non-convexity indicates that these kind of functions have many extreme values so that the system can have several relatively stable equilibrium states. This will lead to diversity in system evolution.

In ANNs, neuron processing units can represent different objects such as characteristics, letters, concepts or some meaningful abstract patterns. In a network, processing units fall mainly into three categories: input units, output units, and hidden units. Input units receive signals and data from outside. Output units output the processing results and hidden units, which are in the middle of input units and output units, cannot be observed from outside the system. The connection weights between neurons reflect the connection strength of units, while the presentation and processing of information are embodied in the connection between processing units. The way in which ANNs conduct information processing is non-procedural, adaptive, and follows a process similar to the human brain. The nature of ANNs is to be equipped with a parallel and distributed information processing function through network transformation and dynamic behaviors. ANNs at the same time simulate the information processing function done by the human nervous system in varying degrees and at different levels. The study of ANNs is interdisciplinary involving many different fields, such as neural science, noetic science, artificial intelligence, computer science, and so on.

ANNs use parallel and distributed systems and adapt a mechanism completely different from traditional artificial intelligence and information processing technology. In this way, ANNs overcome obstacles encountered by traditional logic and symbol-based artificial intelligence in processing intuitive and unstructured information. In addition, ANNs have the characteristics of self-adaption, self-organization, and real-time learning.

The ANN is an important research direction in the field of artificial intelligence and has three advantages. First, it has the function of self-learning. For example, image recognition can be easily done if a large number of different image templates and their corresponding recognition results have been inputted into the artificial neural network. This is because the network can gradually learn to recognize similar images through its self-learning function. Second, ANNs have associative memories which can be achieved by means of using ANNs' feedback networks. Third, ANNs are able to find the optimal solution at high speed. A large computational workload is often required to produce the optimal solution to a complex problem. The computing time can be greatly reduced by using a high-speed feedback-type ANN designed specifically for the problem.

5.1.3 Basic Principles of Neural Networks

No matter what kind of neural network model is discussed, the minimum information processing unit is the neuron. So far, people have built hundreds of artificial neuron models. However, the most commonly used is still the earliest MP Model. The neuron is a multi-input and single-output information processing unit and a neural network which consists of several neurons through a weighted connection. Though a single neuron is only able to do very simple information processing, a network connected by more than one neuron has stronger computing capability. Neural network computing manifests as the interaction between neurons. By changing the connection mode and strength between neurons, the computational efficiency of neural networks can be changed. The connection strength between two neurons is denoted by a real number, which is called the connection weight. The connection form and connection weight between neurons are often determined by the learning process of neural networks. Based on different types, connection modes and learning styles of neurons, various neural network models are designed. The structure of neural networks is shown in Fig. 5.1.

ANN models mainly focus on the topological structures in network connections, the characteristics of neurons and the learning rules. Currently, there are nearly 40 neural network models, including the BP Network, the self-organizing map, the Hopfield network, the Boltzmann machine network, the Adaptive resonance theory network, and so on. According to the topological structures in network connections, neural network models can be divided as follows:

① Forward networks. In a forward network, each neuron inputs the information from the previous level and outputs the information to the next level without any feedback and can be represented by a directed acyclic graph. This kind of network can achieve the transformation of signals from the input space to the

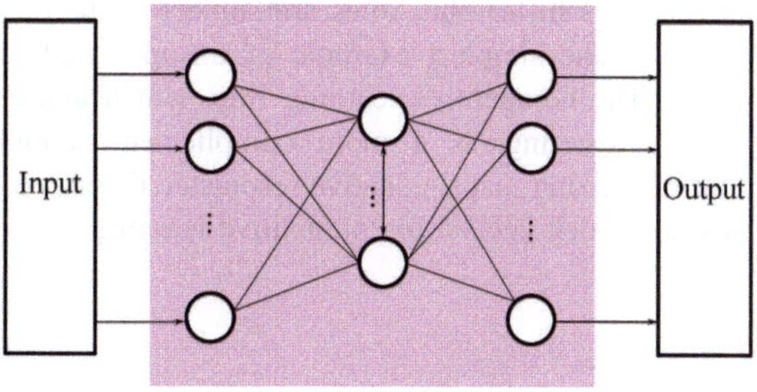

Fig. 5.1 Structure of neural networks

output space. Its information processing ability comes from the multiple compositions of simple nonlinear functions. Thanks to the simple network structure, it is easy to create a forward network. The BP Network is a typical forward network.

② Feedback networks. In a feedback network, feedbacks exist between neurons and the working process can be described by an undirected complete graph. The information processing of this kind of neural network is actually the transformation of states and can be treated by using a dynamic system theory. The stability of the system is closely related to the associative memory function. Both Hopfield Model and Boltzmann Machine belong to this kind of network.

Learning is an important topic in neural network research. The adaptability of neural networks is obtained through learning. Based on environmental changes, the weights are adjusted accordingly in order to improve the performance of the system. The Hebb Learning Rules proposed by Hebb lays the foundation for the learning algorithms of neural networks. Hebb Learning Rules hold that the learning process ultimately happens at the synapses between neurons. The connection strength of synapses changes with the activities of neurons around synapses. It is on this basis that people have proposed various learning rules and algorithms in order to meet the demands of different network models. Efficient learning algorithms enable the neural networks to formulate the intrinsic representation of the objective world and to establish featured information processing methods by adjusting connection weights. The storage and processing of information are reflected in the connection of networks.

According to different learning environments, the learning methods of neural networks can be divided into supervised learning and unsupervised learning. In the process of supervised learning, data of training samples is placed at the input end and at the same time, by comparing the expected output with the network output, error signals can be obtained. Based on this, the connection strength of weights is adjusted. After several trainings, a determined weight is obtained by convergence. When samples change, weights can be modified through learning in order to adapt to a new environment. Neural networks using supervised learning include back propagation networks, perceptrons, and so on. In the process of unsupervised learning, the network is directly placed in a new environment without giving a standard sample. The learning stage and working stage are integrated. At this time, the changes in learning rules are subject to the evolution equation of connection weights. The simplest example of unsupervised learning is the Hebb Learning Rules. The competitive learning rule, which adjusts weights according to established clustering, is a more complicated example of unsupervised learning. Self-organizing maps, adaptive resonance theory networks and many others are all typical models related to competitive learning.

5.2 Star Identification Utilizing Neural Networks Based on Features of a Star Vector Matrix

The star identification algorithm carried out by using neural networks based on features of star vector matrix [5] makes use of the direction vectors of one primary star and three other stars in the neighborhood in order to establish the primary star's feature vector, which is considered as the weight vector of a self-organizing competitive network. Star identification is finished "automatically" by using the competition mechanism of self-organizing competitive networks.

5.2.1 Self-organizing Competitive Neural Networks

In actual neural networks, such as the human retina, there is "lateral inhibition," which means if one neuron is excited, it will inhibit other neurons around through its branches. Lateral inhibition brings out competition between neurons. Though at the initial stage, every neuron is in different levels of excitation, lateral inhibition makes neurons compete with each other. Finally, the inhibition produced by the neuron with the strongest excitatory effect defeats the inhibitions produced by other neurons. So this neuron "wins" and neurons around the "winner" are all "losers."

Self-organizing competitive neural networks are formed on the basis of the above-mentioned biological structure and phenomenon. These kinds of networks can conduct self-organizing training and judgments to input patterns and ultimately divide these patterns into different types. In structure, self-organizing competitive neural networks are often single layer networks consisting of an input layer and a competitive layer. There is not a hidden layer and neurons between the input layer and the competitive layer connect bidirectionally. At the same time, neurons in the competitive layer also have transverse connections. The basic idea, here is that in the competitive layer, neurons compete to respond to the input pattern and only one neuron will be the final winner. In addition to the competition, neurons can also become a winner by producing inhibitions. That means every neuron can inhibit other neurons from responding and make itself the winner. Moreover, there is another lateral inhibition method by which every neuron only inhibits its neighboring neurons but not neurons far away. In learning algorithms, the network simulates the dynamics principles of biological nervous systems which conduct information processing by excitation, coordination, inhibition, and competition between neurons to supervise its learning and work. Therefore, the self-organizing and self-adaptive learning ability of self-organizing competitive neural networks further broadens the application of neural networks in pattern recognition and classification.

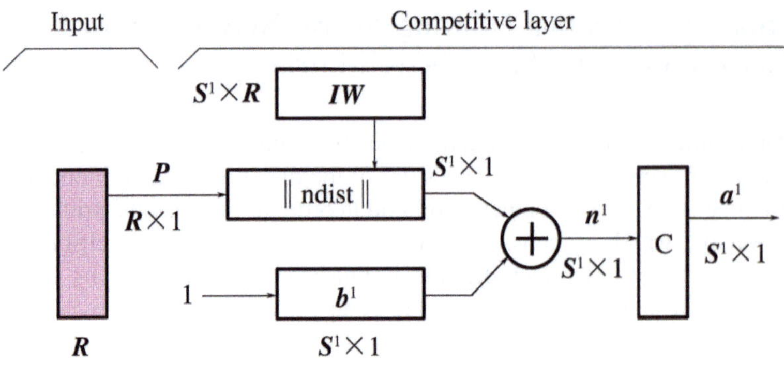

Fig. 5.2 Structural diagram of self-organizing competitive neural networks

Self-organizing competitive neural networks are shown in Fig. 5.2.

In the structural diagram, the inputs of ‖ndist‖ are input vector P and weight matrix IW. The output is an $S^1 \times 1$-dimension vector, which represents the negative distance between the input vector and weight row vector. The algorithm is as follows:

$$\|\text{ndist}\| = -\|\text{IW}_i - \text{P}\|$$

i stands for the ith weight vector in the weight matrix IW.

The network input data n^1 in the competitive layer is the sum of the negative distance and threshold b^1 and is an $S^1 \times 1$-dimension vector. If the entire threshold vector is 0, when input vector P and weight vector IW are equal, n^1 reaches its maximum value 0.

For the biggest element in n^1, the output of the transfer function in the competitive layer is 1, while its output is 0 for other elements. If all the thresholds are 0, the neuron whose weight vector is the closest to the input vector has the minimum negative value but the maximum absolute value. Thus, this neuron wins the competition and the output result is 1.

Star identification utilizing self-organizing competitive neural networks has the following advantages:

① Simple in structure. This kind of network has only two layers of structure without a hidden layer. So it is easy to understand and compute.

② Easy to train. Each star is an independent pattern. In the process of star identification, learning clustering is not needed. Under this circumstance, there is no need for iterative computing weights.

③ Clear in results. The nodes of the output layer are either 0 or 1 and can clearly indicate which star is represented by the identification result. However, some other networks output a real number between 0 and 1. They can indicate the class of the star only after adjustments.

5.2.2 Extraction and Storage of Guide Star Patterns

When formulating the patterns, a guide star is chosen as the primary star and three stars around it are chosen as neighboring stars. All the four stars consist of the guide star pattern. As to different guide stars, their three neighboring stars also have different positions, so the pattern of a particular guide star is unique. The direction vectors of these four stars in the celestial coordinate system can form a vector matrix V. According to the characteristics of star sensor imaging models, after transposing V and multiplying it by the original matrix, a symmetrical characteristic matrix $V^{\mathrm{T}}V$ can be obtained. Similarly, four corresponding measured stars of these four guide stars can also form a symmetric matrix $W^{\mathrm{T}}W$. $V^{\mathrm{T}}V$ and $W^{\mathrm{T}}W$ are completely identical. This means no matter whether in the celestial coordinate system or the image coordinate system, the symmetric matrix formed by the same set of stars remains unchanged. So this symmetric matrix can be viewed as a characteristic for star identification.

Denoting the right ascension and declination coordinates of the guide star as (α, δ), its direction vector in the celestial coordinate system is $[\cos\alpha\cos\delta \quad \sin\alpha\cos\delta\ \sin\delta]^{\mathrm{T}}$. The transformation of the vector of stars from the celestial coordinate system into the star sensor coordinate system is $W = AV$. Here W stands for the direction vector matrix of the measured star in the star sensor coordinate system, A for the attitude matrix and V for the direction vector matrix of the guide star in the celestial coordinate system. When four stars are observed, $W = [b_1 \quad b_2 \quad b_3 \quad b_4]$, $V = [r_1 \quad r_2 \quad r_3 \quad r_4]$. Here b_i is the direction vector of the measured star, r_i is the direction vector the guide star, i = 1, 2, 3, 4. W is a 3 × 4 matrix and $W^{\mathrm{T}}W$ is a 4 × 4 matrix.

Because $W = AV$, it follows that

$$[b_1 \quad b_2 \quad b_3 \quad b_4] = A[r_1 \quad r_2 \quad r_3 \quad r_4] \tag{5.1}$$

$$\begin{aligned} W^{\mathrm{T}}W &= [b_1 \quad b_2 \quad b_3 \quad b_4]^{\mathrm{T}}[b_1 \quad b_2 \quad b_3 \quad b_4] \\ &= \begin{bmatrix} b_1^{\mathrm{T}}b_1 & b_1^{\mathrm{T}}b_2 & b_1^{\mathrm{T}}b_3 & b_1^{\mathrm{T}}b_4 \\ b_2^{\mathrm{T}}b_1 & b_2^{\mathrm{T}}b_2 & b_2^{\mathrm{T}}b_3 & b_2^{\mathrm{T}}b_4 \\ b_3^{\mathrm{T}}b_1 & b_3^{\mathrm{T}}b_2 & b_3^{\mathrm{T}}b_3 & b_3^{\mathrm{T}}b_4 \\ b_4^{\mathrm{T}}b_1 & b_4^{\mathrm{T}}b_2 & b_4^{\mathrm{T}}b_3 & b_4^{\mathrm{T}}b_4 \end{bmatrix} = \begin{bmatrix} 1 & b_1^{\mathrm{T}}b_2 & b_1^{\mathrm{T}}b_3 & b_1^{\mathrm{T}}b_4 \\ b_2^{\mathrm{T}}b_1 & 1 & b_2^{\mathrm{T}}b_3 & b_2^{\mathrm{T}}b_4 \\ b_3^{\mathrm{T}}b_1 & b_3^{\mathrm{T}}b_2 & 1 & b_3^{\mathrm{T}}b_4 \\ b_4^{\mathrm{T}}b_1 & b_4^{\mathrm{T}}b_2 & b_4^{\mathrm{T}}b_3 & 1 \end{bmatrix} \end{aligned} \tag{5.2}$$

$$W^{\mathrm{T}}W = V^{\mathrm{T}}A^{\mathrm{T}}AV = V^{\mathrm{T}}V \tag{5.3}$$

$$\begin{aligned} V^{\mathrm{T}}V &= [r_1 \quad r_2 \quad r_3 \quad r_4]^{\mathrm{T}}[r_1 \quad r_2 \quad r_3 \quad r_4] \\ &= \begin{bmatrix} r_1^{\mathrm{T}}r_1 & r_1^{\mathrm{T}}r_2 & r_1^{\mathrm{T}}r_3 & r_1^{\mathrm{T}}r_4 \\ r_2^{\mathrm{T}}r_1 & r_2^{\mathrm{T}}r_2 & r_2^{\mathrm{T}}r_3 & r_2^{\mathrm{T}}r_4 \\ r_3^{\mathrm{T}}r_1 & r_3^{\mathrm{T}}r_2 & r_3^{\mathrm{T}}r_3 & r_3^{\mathrm{T}}r_4 \\ r_4^{\mathrm{T}}r_1 & r_4^{\mathrm{T}}r_2 & r_4^{\mathrm{T}}r_3 & r_4^{\mathrm{T}}r_4 \end{bmatrix} = \begin{bmatrix} 1_1 & r_1^{\mathrm{T}}r_2 & r_1^{\mathrm{T}}r_3 & r_1^{\mathrm{T}}r_4 \\ r_2^{\mathrm{T}}r_1 & 1 & r_2^{\mathrm{T}}r_3 & r_2^{\mathrm{T}}r_4 \\ r_3^{\mathrm{T}}r_1 & r_3^{\mathrm{T}}r_2 & 1 & r_3^{\mathrm{T}}r_4 \\ r_4^{\mathrm{T}}r_1 & r_4^{\mathrm{T}}r_2 & r_4^{\mathrm{T}}r_3 & 1 \end{bmatrix} \end{aligned} \tag{5.4}$$

Because b_i and r_i are unit vectors, diagonal elements of the matrix in both Eqs. (5.1) and (5.4) are 1. Therefore based on Eqs. (5.2), (5.3), and (5.4), the following equation can be obtained

$$\begin{bmatrix} 1 & b_1^{\mathrm{T}}b_2 & b_1^{\mathrm{T}}b_3 & b_1^{\mathrm{T}}b_4 \\ b_2^{\mathrm{T}}b_1 & 1 & b_2^{\mathrm{T}}b_3 & b_2^{\mathrm{T}}b_4 \\ b_3^{\mathrm{T}}b_1 & b_3^{\mathrm{T}}b_2 & 1 & b_3^{\mathrm{T}}b_4 \\ b_4^{\mathrm{T}}b_1 & b_4^{\mathrm{T}}b_2 & b_4^{\mathrm{T}}b_3 & 1 \end{bmatrix} = \begin{bmatrix} 1_1 & r_1^{\mathrm{T}}r_2 & r_1^{\mathrm{T}}r_3 & r_1^{\mathrm{T}}r_4 \\ r_2^{\mathrm{T}}r_1 & 1 & r_2^{\mathrm{T}}r_3 & r_2^{\mathrm{T}}r_4 \\ r_3^{\mathrm{T}}r_1 & r_3^{\mathrm{T}}r_2 & 1 & r_3^{\mathrm{T}}r_4 \\ r_4^{\mathrm{T}}r_1 & r_4^{\mathrm{T}}r_2 & r_4^{\mathrm{T}}r_3 & 1 \end{bmatrix} \tag{5.5}$$

In Eq. (5.5), corresponding elements on two sides of the equation are equal. Because they are all symmetric matrixes, $b_1^{\mathrm{T}}b_2 = r_1^{\mathrm{T}}r_2$, $b_1^{\mathrm{T}}b_3 = r_1^{\mathrm{T}}r_3$, $b_1^{\mathrm{T}}b_4 = r_1^{\mathrm{T}}r_4$, $b_2^{\mathrm{T}}b_3 = r_2^{\mathrm{T}}r_3, b_2^{\mathrm{T}}b_4 = r_2^{\mathrm{T}}r_4$, and $b_3^{\mathrm{T}}b_4 = r_3^{\mathrm{T}}r_4$. In other words, for any set of stars in the celestial coordinate system, if any two of their direction vectors are multiplied by each other, the products remain unchanged when the vectors are converted into the star sensor coordinate system. So these products can be extracted and used as characteristics for star identification. As four stars are selected to form feature vectors, six elements in the matrix formed according to Eqs. (5.2) and (5.3) are independent. These six elements can be used to form star's feature vectors as follows:

The feature vector of the guide star is

$$pat_b = \left[b_1^{\mathrm{T}}b_2 \quad b_1^{\mathrm{T}}b_3 \quad b_1^{\mathrm{T}}b_4 \quad b_2^{\mathrm{T}}b_3 \quad b_2^{\mathrm{T}}b_4 \quad b_3^{\mathrm{T}}b_4 \right].$$

The feature vector of the measured star is

$$pat_r = \left[r_1^{\mathrm{T}}r_2 \quad r_1^{\mathrm{T}}r_3 \quad r_1^{\mathrm{T}}r_4 \quad r_2^{\mathrm{T}}r_3 \quad r_2^{\mathrm{T}}r_4 \quad r_3^{\mathrm{T}}r_4 \right].$$

This feature vector is formed by four stars and contains the relative positions of these four stars, so it can be viewed as the pattern of these four stars. Moreover, if one of the stars is selected to be the primary star, this feature vector can also reflect the distribution of the neighbor stars. So the feature vector can also be considered as the primary star's pattern.

To form guide star patterns, every star in the GSC is selected as the primary star in turn and three neighboring stars in the neighborhood of the primary star are found. These four stars are considered as a group to form the feature vector of this set of stars according to $pat_b = \left[b_1^{\mathrm{T}}b_2 \quad b_1^{\mathrm{T}}b_3 \quad b_1^{\mathrm{T}}b_4 \quad b_2^{\mathrm{T}}b_3 \quad b_2^{\mathrm{T}}b_4 \quad b_3^{\mathrm{T}}b_4 \right]$. The reason for choosing three neighboring stars around the primary star to structure a feature vector is that there are no two primary stars whose three neighbor stars are in exactly the same pattern distribution. So three neighbor stars, as well as one primary star, are enough to form the feature vector.

When forming measured star patterns, if the primary star is far away from the image center, then it is possible that the neighbor stars may fall outside the image. This makes neighbor stars around the primary star incomplete. Meanwhile, the more neighboring stars are selected, the larger probability that neighbor stars are mistakenly selected.

As to different primary stars, their three neighbor stars have different positions, and their corresponding feature vectors are also different. According to the rule of the pattern class identification algorithm, the geometric distribution patterns of neighboring stars in a certain neighborhood can form a unique pattern of the primary star. So, this feature vector can be used as the pattern of the primary star.

The procedures of forming the feature vector of the primary star are as follows:

① Determine the neighboring stars to be selected. For any primary star S_1, compute the angular distances R_i between the primary star and all the neighboring stars around it. Here i = 1, 2, 3… n, and n stands for the total number of neighboring stars around the primary star. Take stars whose angular distances are in the range of $R_t < R_i < R_{\mathrm{FOV}}$ to be the neighboring stars to be selected. Here, R_{FOV} is the FOV of star sensors, and R_t is in the segment 0.5°–1°.

② Determine the neighbor stars to form the feature vector. According to the angular distances between the primary star S_1 and the neighboring stars to be selected, order all the neighboring stars to be selected from the smallest to the largest. Select three stars S_2, S_3, and S_4 closest to the primary star S_1 to be the neighboring stars to form the feature vector.

③ Form the feature vector of the primary star S_1. Compute the direction vectors b_1, b_2, b_3, b_4 between the primary star S_1 and three neighboring stars S_2, S_3, S_4. The algorithm is as follows:

$$b_i = \begin{bmatrix} \cos\alpha_i \cos\delta_i & \cos\alpha_i \sin\delta_i & \sin\delta_i \end{bmatrix}^{\mathrm{T}}.$$

Here, b_i stands for the direction vectors of the primary star S_1 and three neighbor stars S_2, S_3, S_4, i = 1, 2, 3, 4.

In this equation, α_i stands for the right ascension of the ith star and δ_i stands for its declination.

Compute and form the feature vector $pat_b = \begin{bmatrix} b_1^{\mathrm{T}}b_2 & b_1^{\mathrm{T}}b_3 & b_1^{\mathrm{T}}b_4 & b_2^{\mathrm{T}}b_3 & b_2^{\mathrm{T}}b_4 & b_3^{\mathrm{T}}b_4 \end{bmatrix}$ of the primary star S_1. This feature vector is a six dimensional vector. Then, compute and store all the feature vectors of all the guide stars in the GSC.

In addition to storing the feature vectors of guide stars, these guide stars used to form these patterns also need to be stored. This is because when a primary star is obtained, it is easy to look up the indexes of three neighboring stars used to form the pattern with the primary star. In the operation of self-organizing competitive neural networks, only one node outputs 1 and only the index of the primary star can be obtained. The indexes of three neighboring stars around the primary star cannot be determined. By looking up the data in the index database of neighboring stars, the neighbor stars around the primary star can be found. And in the process of identification, the corresponding guide star of the neighboring stars in the measured star image can be quickly found. In other words, the three neighbor stars can be identified.

The storage format of a partial record of the neighbor star index database is shown in Fig. 5.3.

Fig. 5.3 Storage format of neighbor star index database

Primary star	Neighbor star1	Neighbor star2	Neighbor star3
24	3	32	29
25	27	26	28
…	…	…	…

5.2.3 Construction of Self-organizing Competitive Neural Networks

Self-organizing competitive neural networks used for star identification have two layers. The first layer is the input layer, in which the number of nodes is the same as the number of dimensions of the feature vector of the guide star. That is, there are six nodes. The second layer is the output layer. The number of nodes is the same as that of classes. The number of the class number is the total number of guide stars. Every node in the output layer is connected with one node in the input layer. The weights between the nodes of the two layers are called weight vectors.

When the self-organizing competitive neural networks are constructed, the values of the weight vectors between all the nodes are determined according to the classification. This is the so-called network training. The typical learning rule adopted in competitive learning strategies is Winner-Takes-All. The principle of competitive learning is shown in Fig. 5.4. Denoting the input pattern as a two dimensional vector. After normalization, its vector end can be viewed as a spot of identity element distributed in the image and is represented by "O." Denoting there are three neurons in the competitive layer and that their three corresponding star vectors are marked in the same unit circle. After been fully trained, the three "★"

Fig. 5.4 Illustration of competitive learning principle

spots in the unit circle gradually shift into the cluster center of every input feature vectors. Thus, the weight vectors of all the neurons of the competitive layer become the clustering centers of the input feature vector. When a pattern is inputted into the network, the winning neuron in the competitive layer outputs 1 and the input is put into the same class as the winner belongs to.

In star identification, every guide star is viewed as a class of the output layer and every class has only one feature pattern. So, after normalizing the feature patterns of all guide stars, weight vectors of corresponding nodes in the output layer are assigned directly. Thus, the training of self-organizing competitive neural networks is completed. As is shown in Fig. 5.5, every class has only one feature vector, or one "O" spot. The position of "O" is the clustering center of this class of feature vectors. "★" represents the position of the weight vector of a node. Move "★" to all the "O" to make them overlapping, so the weight vectors of all the classes point to their own clustering centers. The values of weight vectors and the input feature patterns are equal.

In the training process of networks, the feature vectors of guide stars are used to train the self-organizing competitive neural networks. The pattern information of guide stars is integrated into the weight matrix of neural networks. So in well-trained networks, patterns of all the guide stars are included. In the process of identification, there is no need to read the data of guide star patterns and store the information individually.

In measured star images, the primary star to be identified and three neighboring stars used to form the pattern are determined. Then the feature vector of the primary star to be identified is formed. Input the feature vector into the well-trained self-organizing competitive neural network. The network judges which node is the closest to the input pattern and the corresponding node in the output layer outputs 1.

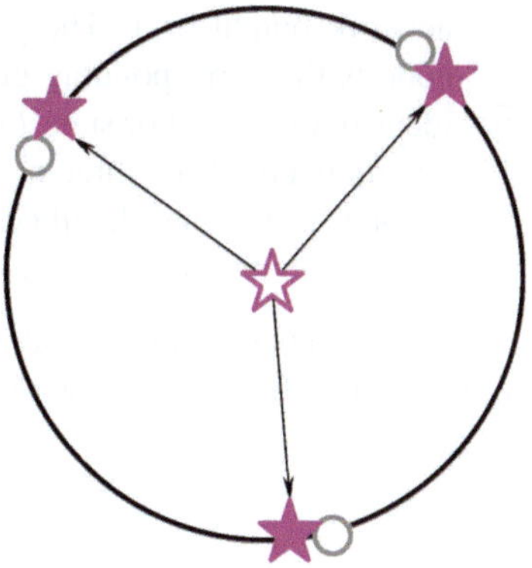

Fig. 5.5 Training principle of self-organizing competitive neural networks used for star identification

Look up the index number of the node. This number is the corresponding guide star index of the primary star to be identified.

① Determine the primary star to be identified T_1. When a measured star image is obtained, compute the distances between all the stars and the central point of the image. Then sort the stars according to the distances in ascending order and select the star with the minimal distance as the primary star to be identified T_1.

② Determine the neighboring stars used to form the feature vector. Compute the angular distances between the primary star to be identified T_1 and the neighboring stars around it. Select three neighboring stars T_2, T_3, T_4 with the minimal angular distances as the neighboring stars used to form the feature vector.

③ Form the feature vector of the primary star to be identified T_1. Compute the direction vectors r_1, r_2, r_3, r_4 of the primary star to be identified T_1 and its neighbor stars T_2, T_3, T_4 according to the algorithm as follows:

$$r_i = \left[\frac{x_i}{\sqrt{x_i^2+y_i^2+f^2}} \quad \frac{y_i}{\sqrt{x_i^2+y_i^2+f^2}} \quad \frac{f}{\sqrt{x_i^2+y_i^2+f^2}} \right]^T,$$

r_i stands for the direction vectors of the primary star to be identified T_1 and its neighbor stars T_2, T_3, T_4, $i = 1, 2, 3, 4$.

Here, x_i is the horizontal coordinate of the measured star in the image plane, y_i is the vertical coordinate of the measured star in the image plane, f is the focal length of the star sensors.

Form the feature vector of the primary star to be identified T_1:

$$pat_r = \left[r_1^T r_2 \quad r_1^T r_3 \quad r_1^T r_4 \quad r_2^T r_3 \quad r_2^T r_4 \quad r_3^T r_4 \right].$$

④ Identify the primary star to be identified. Input pat_r into the well-trained self-organizing competitive neural network and look up the node with the network output of 1. The guide star determined by the index number of the node is the corresponding guide star of the primary star to be identified T_1.

⑤ Identify the neighbor stars T_2, T_3, T_4. In the neighbor star database, look up the star numbers of the three neighbor stars of the corresponding guide star of the primary star to be identified T_1. These three neighbor stars are in correspondence with neighbor stars T_2, T_3, T_4.

The flow chart of the neural network star identification algorithm by using star vector matrix feature is shown in Fig. 5.6.

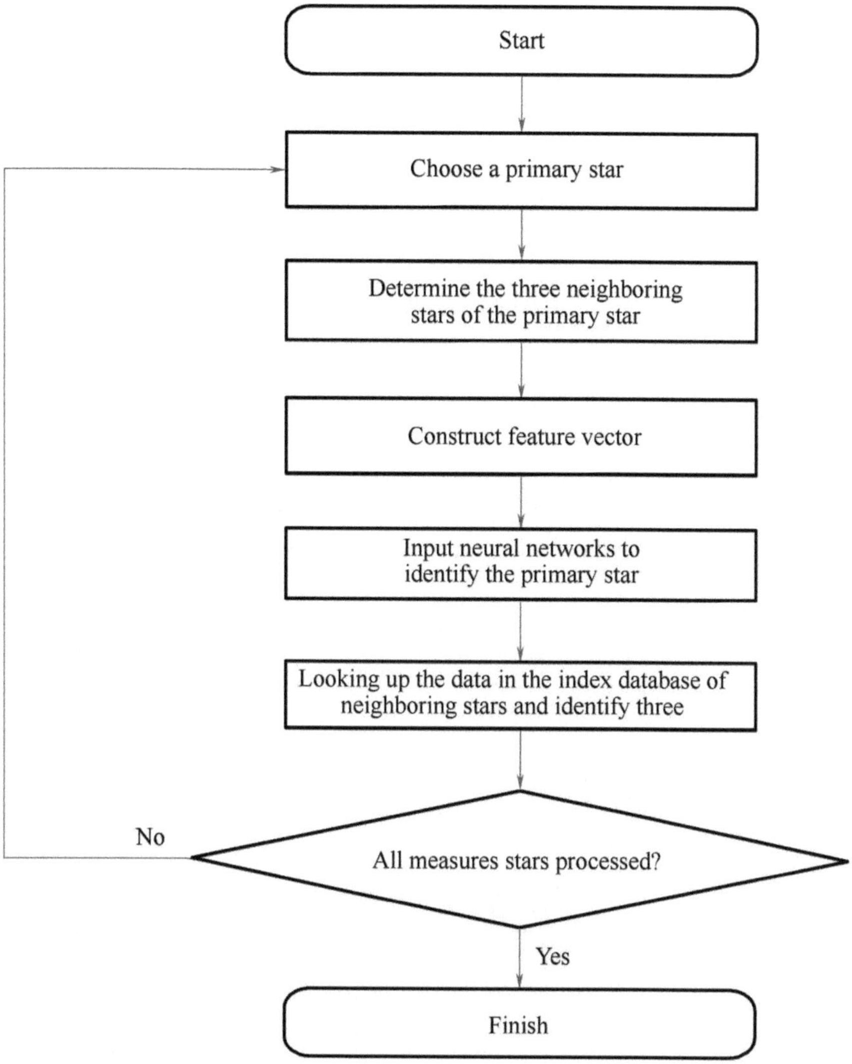

Fig. 5.6 Flow chart of the neural network star identification algorithm by using the star vector matrix feature

5.2.4 Simulations and Results Analysis

In simulations, the imaging parameters of star sensors are as follows: the FOV size is 10.8° × 10.8°, the focal length of the optical system is 80.047 mm, the pixel size is 0.015 mm × 0.015 mm, and the pixel resolution is 1024 × 1024. Select stars brighter than 6 Mv from the SAO J2000 Fundamental Star Catalog to form a GSC and generate a corresponding pattern vector database. The simulations are realized on Intel Pentium 4 2.0 GHz computers with MATLAB.

(1) Example of Identification

Figure 5.7 is the identification result of four random star images, + stands for measured stars in the FOV and O stands for stars being correctly identified. It can be seen that, similar to other methods of star identification based on "star patterns," the probability of correctly identifying stars close to the center of the FOV is higher than that of stars near the image edge of the FOV. Stars close to the image edge have a larger probability of having their neighboring stars missing. Missing stars lead to incomplete patterns and consequently may make the stars fail to be identified correctly.

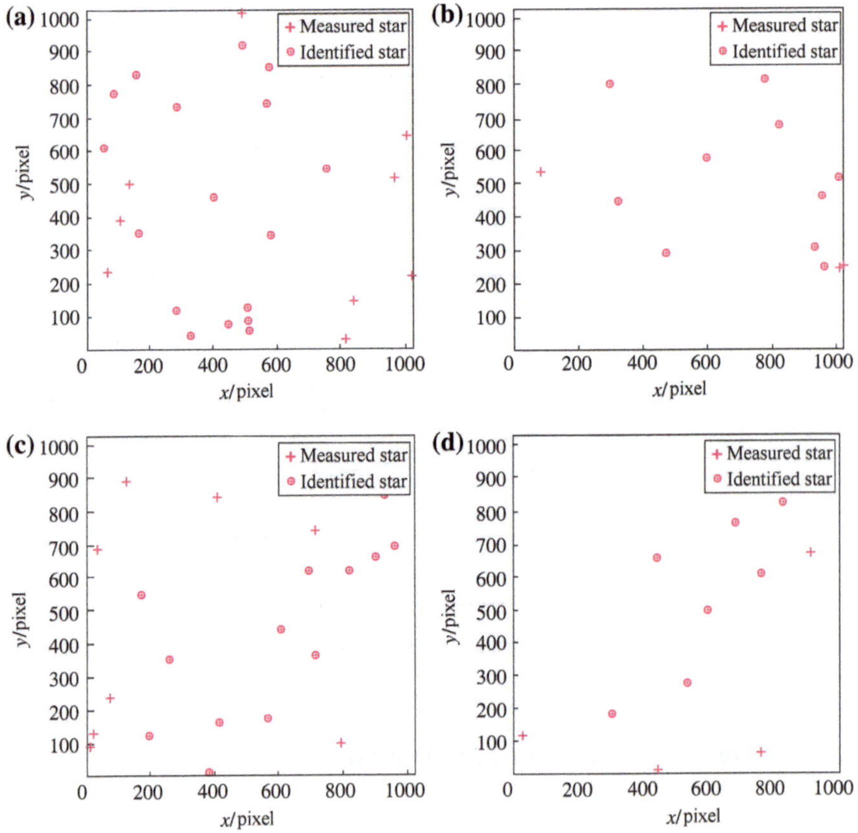

Fig. 5.7 Identification result of four random star images

(2) Impact of Position Noises on Identification Rates

In star images generated through simulation, add Gaussian Random Noise with a mean value of 0 and a standard deviation from 0 to 2 pixels to the star spot, in order to study the identification rates of the algorithm under the impact of position noises. Under various levels of noise, 1000 random star images are generated, respectively, and identified by this method one by one. The identification results of these simulated star images are studied and compared with those obtained through the grid algorithm.

From Fig. 5.8, it can been seen that as the position noises increase, though the identification rates of both two methods decrease, comparatively high identification performances are maintained on the whole. Even when the standard deviation is 2 pixels, the identification rate can still reach 96%. Under various levels of noises, the identification rate of this algorithm is slightly higher than that of the grid algorithm.

(3) Impact of Magnitude Noises on Identification Rates

In order to study the performances of identification algorithms when there are magnitude noises, noises with a mean value of 0 and standard deviation from 0 to 1 Mv are added to the generated star images. When magnitude noises of different levels are added, the identification rates of two algorithms are studied and the result is shown in Fig. 5.9. This is obtained by identifying 1000 randomly generated images from the whole celestial sphere. As the level of magnitude noises increases, the identification rates of two algorithms are rarely affected. Since information about star magnitude is not used in forming the pattern, magnitude noises have little impact on identification rates.

Fig. 5.8 Impact of position noises on identification rates

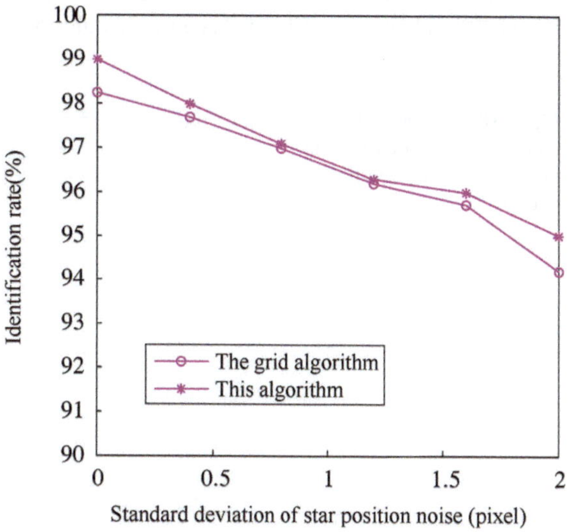

Fig. 5.9 Impact of magnitude noises on identification rates

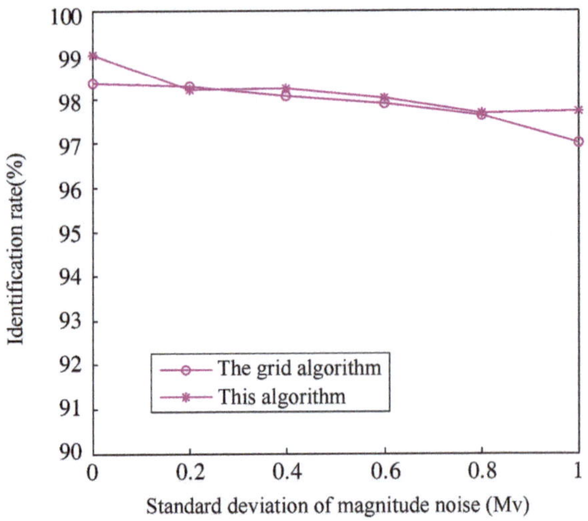

(4) Impact of the Number of Measured Stars in the FOV on Identification Rates

For algorithms based on star pattern classes, the number of measured stars in the FOV is an important factor that can influence identification performances. The more measured stars in the FOV, the easier to form unique patterns of stars and to do identification. At this time, an algorithm based on pattern class is often outstanding in performance. From Fig. 5.10, it can be seen that when the star number in the FOV reaches 10, this algorithm and the grid algorithm both have identification rates of 95% or above. When the star number exceeds 12, both algorithms obtain identification rates of nearly 100%. When the star number is under 9, this algorithm has greater advantages. This is because this algorithm only selects the primary star

Fig. 5.10 Impact of the number of measured stars in the FOV on the identification rates

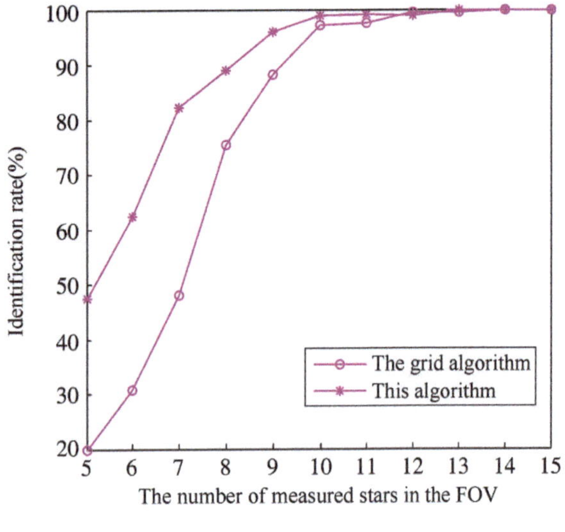

and the three neighboring stars around to generate the star pattern. However, the grid algorithm divides the image and the resolution decreases consequently. So when there are not enough stars, the grid algorithm cannot describe the pattern of the primary star accurately and the identification rate drops sharply.

(5) Identification Time and Storage Capacity

This algorithm simulates in a MATLAB environment and the average identification time for one star image is 0.4 s. The networks store the patterns of the guide stars in weight vectors and make them a part of the network. So there is no need to store the patterns of the guide stars separately. In addition, the GSC and the neighbor star index database are needed when running the algorithm. The two altogether require about 280 KB.

5.3 Star Identification Utilizing Neural Networks Based on Mixed Features

Similar to 5.1, the star identification algorithm by using neural networks based on mixed features [6] uses mixed features consisting of close neighboring triangles and radial-distributed vectors to form feature patterns of stars. The algorithm uses competitive networks to complete star identification.

5.3.1 Construction of Competitive Neural Networks

Competitive networks can be simplified to the simplest form—the Hamming Network. Under this circumstance, the weight matrix of the networks consists of the pattern vectors of all guide stars.

The structure of the Hamming Network is shown in Fig. 5.11. The network consists of two layers: The first layer is used to compute the dot product (the correlation degree) between an input vector and an original vector. The second layer judges which prototype vector is the nearest to the input vector by using the competition mechanism. Denoting the input vector of the measured star as p, and the prototype vectors of the neural network (feature patterns that can be identified by networks) as $\{p_1, p_2, \ldots p_S\}$, the weight matrix and migration vector can be expressed as

$$\mathrm{W}^1 = \begin{bmatrix} w_1^{\mathrm{T}} \\ w_2^{\mathrm{T}} \\ \vdots \\ w_S^{\mathrm{T}} \end{bmatrix} = \begin{bmatrix} p_1^{\mathrm{T}} \\ p_2^{\mathrm{T}} \\ \vdots \\ p_S^{\mathrm{T}} \end{bmatrix}, \quad \mathrm{b} = \begin{bmatrix} R \\ R \\ \vdots \\ R \end{bmatrix}$$

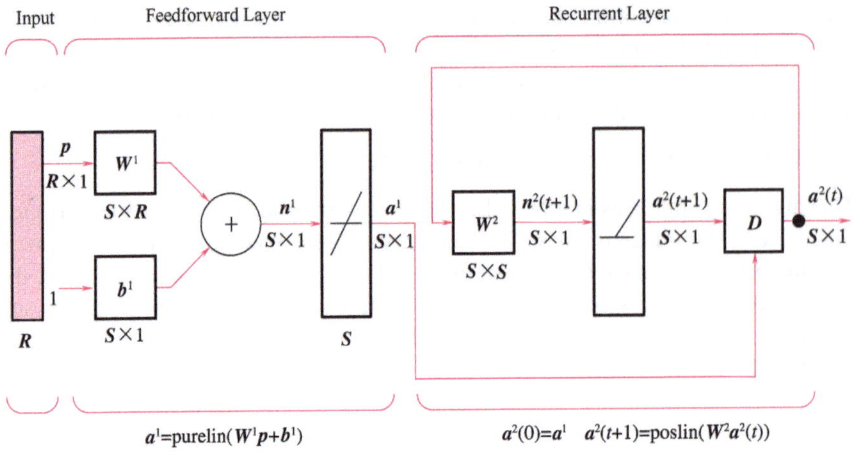

Fig. 5.11 Illustration of hamming network

The output of the first layer is

$$a^1 = W^1 p + b^1 = \begin{bmatrix} p_1^T p + R \\ p_2^T p + R \\ \vdots \\ p_S^T p + R \end{bmatrix} \tag{5.6}$$

The output of the first layer is the initial value of the second layer, and that is

$$a^2(0) = a^1 \tag{5.7}$$

The output of the second layer meets the following recursive procedure

$$a^2(t+1) = \mathrm{poslin}(W^2 a^2(t)) \tag{5.8}$$

Here, W^2 is a matrix with a diagonal of 1 and all other elements of a very small minus ($-\varepsilon$). After iteration, the network ultimately moves toward a stable state. That means the node with the biggest initial value outputs 1 while other nodes output 0. The corresponding prototype vector of the node is the optimal match of the input vector p. So the guide star represented by this node is the matching star of the measured star.

Star identification by using competitive networks has the following advantages:

① A fast and accurate search capability in order to find the optimal match for the input pattern vector. When the input pattern contains noises or is incomplete, the competitive networks can still quickly find the prototype most similar to the input pattern in the pattern space.

② Easy implementation by parallel processing. The structure of competitive networks is fit for parallel computing. If parallel computing is realized with specific integrated circuits, the identification speed will be greatly improved.

5.3.2 Extraction and Identification of Star Pattern Vectors

Generally speaking, patterns fit for star identification by using neural networks are required to meet the following conditions:

① Simple in structure and convenient in computing. Neural networks have to be easy for parallel implementation and generally do not use complicated feature extraction methods.
② Stable and reliable. Patterns must be free of influences from other factors to the greatest extent. For example, when forming the pattern vectors by using the distribution of companion stars in the neighborhood, patterns with rotation invariance are preferred and information on brightness should not be used as much as possible.
③ The pattern of one star should be distinguished from that of other stars. Because every star belongs to a catalog of its own, different stars that have the same or similar patterns should be avoided. This requires using as much information as possible, since one single pattern often fails to distinguish stars from each other.

Through massive experimentation, mixed features consisting of radial patterns and close neighboring triangles are selected to form feature patterns of stars. The structure of feature pattern vector p is shown in Fig. 5.12. It is a 1×13 vector. Denoting the radius of the pattern as r, the definitions of the radial pattern and the close neighboring triangle are as follows:

① Divide the circular neighborhood with radius r into 10 annuluses with the same intervals. Each annulus stands for angular distance $r/10$. R1, R2, …, R10 stand for the number of companion stars falling into the 1st, 2nd, …, 10th annulus,

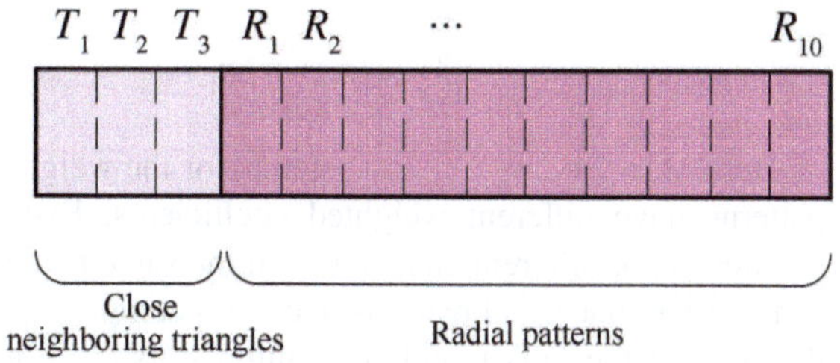

Fig. 5.12 Mixed features

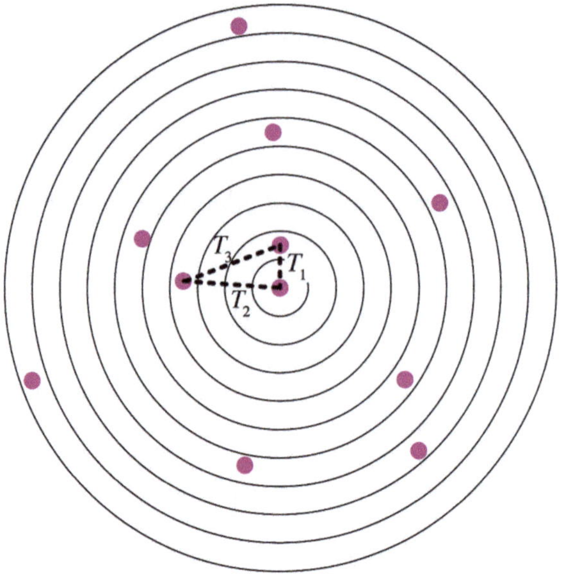

Fig. 5.13 Radial patterns and close neighboring triangles

respectively. The radial pattern vector showed in Fig. 5.13 is (0, 1, 0, 2, 3, 2, 1, 0, 2). Thus, it can be seen that the radial pattern introduced here is different from the radial vector introduced in Sect. 4.2.

② Search the two companion stars closest to the primary star in the area beyond the radius of the neighborhood of the primary star *br* but within radius *r*. T_1 stands for the angular distance between the primary star and its closest companion star, and T_2 for the angular distance between the primary star and its second closest companion star. T_3 stands for the angular distance between the closest companion star and the second closest companion star. T_1, T_2, and T_3 are shown in Fig. 5.13.

As feature vectors consist of mixed features, the radial feature and close neighboring triangle feature have different meanings, the feature vectors cannot be used directly. Different features of the feature vectors should be multiplied by different weighted coefficients, as shown

$$\mathrm{p}' = \mathrm{p} \cdot \mathrm{w} = (\mathrm{T}_1 \cdot w_1, \ldots, \mathrm{R}_{10} \cdot w_{13}) \tag{5.9}$$

Here, $\mathrm{w} = (w_1, w_2, \ldots, w_{13})$ stands for the weighted coefficient vector. Different patterns have different weighted coefficients. Even the corresponding weighted coefficients of different elements with the same patterns are not exactly the same. It can be seen that radial patterns may be incomplete (i.e., measured stars are close to the edge of the FOV), and companion stars close to the primary star have lower probability of being outside the FOV than those companion stars far away. Therefore, closer companion stars are more stable and reliable in redial patterns. This can be reflected in the weighted coefficient vectors as follows:

$$w_4 \geq w_5 \geq \cdots \geq w_{13} \tag{5.10}$$

5.3.3 Simulations and Results Analysis

The algorithm is simulated in the MATLAB environment and the identification results of four randomly generated star images are shown in Fig. 5.14. Here, + stands for the measured stars in the FOV, and O stands for stars correctly identified. It can be seen that, similar to other methods of star identification based on star patterns, the probability of correctly identifying stars close to the center of the FOV is higher than that of stars near the edge of the FOV.

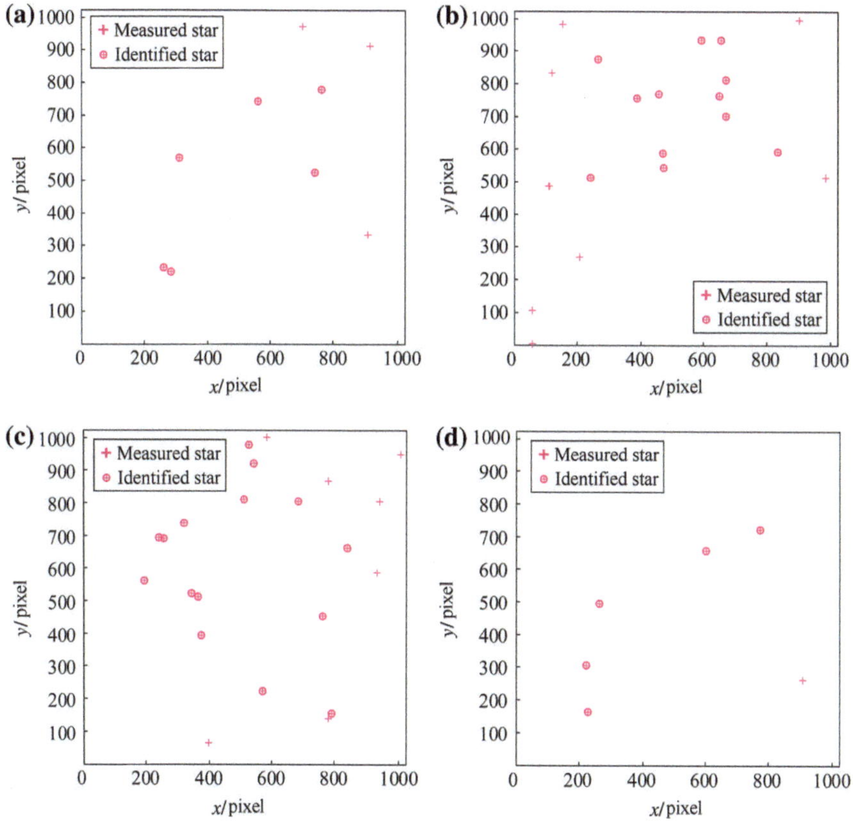

Fig. 5.14 Identification results of four randomly generated star images

Under the circumstance of star spot position noises with standard deviation of 1 pixel and position noises of 0.5 Mv, 1000 randomly generated star images from the whole celestial sphere are identified. The identification rate is 99.7%. This is better than the results for the grid algorithm and the identification algorithm based on radial and cyclic features under the same experimental conditions.

References

1. Lindsey C, Lindblad T (1997) A method for star identification using neural networks. SPIE 3077:471–478
2. Bardwell G (1995) On-board artificial neural network multi-star identification system for 3-axis attitude determination. Acta Astronaut 35:753–761
3. Paladugu L, Schoen M, Williams BG (2003) Intelligent techniques for star-pattern recognition, Proceedings of ASME, IMECE2003-42274
4. Zongli J (2001) Introduction to artificial neural network. High Education Press, Beijing
5. Yang J (2007) Star identification algorithm and application research on RISC technology, Beihang University doctoral dissertation, Beijing, pp 49–60
6. Wei X (2004) Star identification in star sensors and research on correlative technology, Beihang University doctoral dissertation, Beijing, pp 76–81

Chapter 6
Rapid Star Tracking by Using Star Spot Matching Between Adjacent Frames

As described in Sect. 1.4, the star sensor method usually has two working modes, namely the initial attitude establishment mode and the tracking mode. In the initial attitude establishment mode, star sensor identifies and establishes initial attitude by using the full-sky star image. Once the initial attitude of a spacecraft is established successfully, star sensor enters into the tracking mode. In normal conditions, star sensor works in the tracking mode most of the time, which means that the tracking mode is the principle operational mode of star sensor.

In accordance with the star sensor's requirements toward star tracking, Zhang et al. [1–4] put forward a rapid star tracking algorithm by using star spot matching between adjacent frames. The algorithm effectively improves the efficiency of star tracking by taking full advantage of the information of the partition star catalog and using strategies such as threshold mapping, sorting before matching, etc. This chapter introduces the detailed operational process of this algorithm and evaluates its performance through simulation experiments.

6.1 Tracking Mode of the Star Sensor

This section briefly introduces the fundamental principles and process of the tracking mode of star sensor. The characteristics of the star tracking algorithm and star sensor's basic requirements toward star tracking algorithm are analyzed. In the last part, some widely adopted star tracking algorithms are presented.

6.1.1 Principles of Star Tracking

Figure 6.1 demonstrates the operational process of star sensor. As is shown, star sensor has two operational modes.

G. Zhang, *Star Identification*, DOI 10.1007/978-3-662-53783-1_6

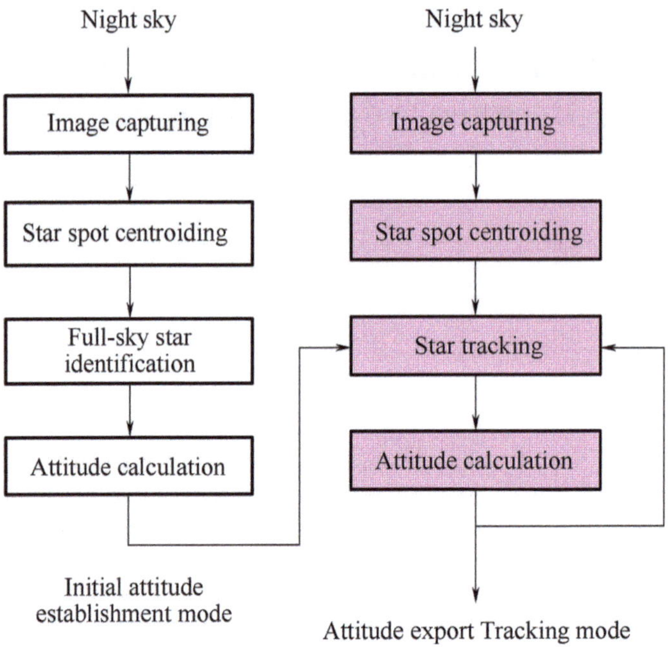

Fig. 6.1 General operation process of star sensor

① Initial attitude establishment mode. With no prior knowledge of the attitude of a spacecraft, star sensor matches and identifies measured star images and then calculates the initial attitude.

② Tracking mode. Based on the results of full-sky star identification and attitude calculation, star sensor uses the identification information in the previous frames of star tracking to track and identify the measured star in the current FOV. Then, attitude can be outputted.

The general operation process of star sensor is as follows:

① When star sensor starts working, it enters into the initial attitude establishment mode and captures star images;

② Precise location of star spots in star images is realized through image processing;

③ The star images are matched and identified all over the celestial sphere, searching for the corresponding guide star of the measured star in star images. The identification results (including the index number of the guide star, star magnitude, right ascension, declination, and other information) are recorded;

④ By calculating the attitude, accurate initial attitude information is obtained;

⑤ Once the initial attitude is available, star sensor enters into tracking mode;
⑥ Star images are captured and star spot centroiding is conducted;
⑦ After tracking, matching and identifying, the results are used to calculate the current attitude of star sensor and the attitude is outputted. Step ⑥ is then repeated for the following rounds of tracking.

6.1.2 Characteristics of Star Tracking

Initial attitude establishment mode and tracking mode are two independent yet correlated working modes of star sensor. Initial attitude establishment mode offers precise initial identification information and initial attitude for the tracking mode. When tracking identification fails or in lost-in-space conditions, star sensor will enter into initial attitude establishment mode and start to identify the full-sky image once again. Lacking initial attitude in initial attitude establishment mode, star sensor carries out star identification regarding the whole celestial sphere as unidentified regions. Hence, a longer time is needed to search and match the stars. The identification usually takes several seconds. In the tracking mode, the results of full-sky star identification and attitude calculation, as well as identification information of the previous frames of star tracking are used for the tracking identification of the measured stars in the current FOV. Therefore, the processing time is relatively short. Only at the initial moment of operation or when faced with a lost-in-space problem, star sensor will enter into initial attitude establishment mode. After the initial attitude is established, star sensor will be in real-time tracking state as long as the tracking mode remains stable. Hence, tracking mode is the major operational mode of star sensor.

Star sensor's requirements for star tracking cover the following aspects:

① Rapidity. The tracking time usually determines the update frequency of the attitude of star sensor. Therefore, the tracking time should be as short as possible.
② Accuracy. The identification results obtained in star tracking mode are directly used for attitude output. Identification errors in the tracking mode may result in fluctuations in attitude output. In serious cases, star sensor may have to start identifying full-sky star images repeatedly.
③ Identify as many measured stars in the FOV as possible. Stars used in attitude calculation are distributed unevenly in the FOV. Hence, if the stars showing up in the FOV are not timely and accurately identified, the attitude export may experience abnormal fluctuations.

6.1.3 Current Star Tracking Algorithms

In terms of specific tracking algorithms, window-based tracking is the most widely adopted one [5–8]. After the identification information of stars and initial attitude has been gained in initial attitude establishment, the next precise position of the tracked star can be obtained through motion vector estimation. A certain form of window is utilized to acquire the image data of star spots, as shown in Fig. 6.2. By using the threshold segmentation approach, star centroids can be obtained and interfering stars are excluded. After star centroiding, if there is one and only one measured star in the tracking window, then this star is considered to be the same measured star that appears in the corresponding position in the last frame. According to the results of star tracking, the current accurate attitude is calculated and used to predict new stars that may appear in the FOV. New stars entering the FOV can be identified by matching four angular distances and then star sensor can proceed to the next round of tracking.

This type of algorithm requires that the image data in sub-windows should be imported into the signal processing unit for the extraction of star centroid position. Restricted by the speed of the interface between image sensor and the signal processing unit, limited star data are transferred. This bottleneck in image data acquisition cannot meet star tracking's requirements for rapidity. Meanwhile, it is demanded by this type of algorithm that real-time information of the position of the sub-window center should be fed back to the image sensor, complicating the driver logic of the image sensor. In addition, these algorithms are expected to estimate precise positional information by utilizing the subsidiary attitude information offered by other inertial devices and filtering with Kalman Filter. The relatively complicated calculation processes of these algorithms are not convenient for fast tracking, moreover, with small tracking windows, attitude disturbances and estimation errors, may result in loss of tracking or acquisition of wrong stars.

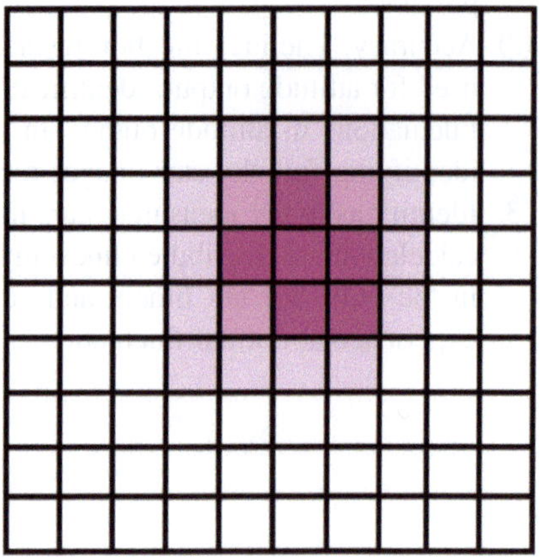

Fig. 6.2 Star tracking

In long-term tracking, it is inevitable that some tracked stars may move out of the tracking FOV and some stars may enter into the FOV. It takes the star sensor produced by Ball Company 0.2 s to accomplish new star identification and attitude estimation. It is thus clear that the identification of new stars is quite time-consuming, which is a bottleneck in rapid star tracking.

6.2 Rapid Star Tracking Algorithm by Using Star Spot Matching Between Adjacent Frames

Rapid star tracking algorithm by using star spot matching between adjacent frames directly uses the corresponding relations of stars between adjacent frames and prior identification information to accomplish the rapid tracking and identification of measured stars in the FOV. In order to accelerate the speed of star tracking, strategies such as zone catalog-based quick retrieval of guide stars, threshold mapping, sorting before tracking and others are used. In this section, the detailed process of the algorithm is presented.

6.2.1 Basic Principles of Star Tracking Algorithm

The general idea of rapid star tracking to be introduced in this chapter is as follows. A reference star image is generated through star prediction. A radius of neighborhood is set to determine if the measured stars in the measured star image is within the neighborhood of the corresponding star in the reference star image. In this way, whether or not the tracking has been accomplished successfully can be evaluated. Through star mapping, the number of tracked stars in the FOV is increased for the convenience of continuous tracking. The process of rapid star tracking is demonstrated in Fig. 6.3.

For better explanation, the meaning and function of each step in the tracking process is briefly explained according to this illustration of star tracking.

1. Initial Attitude

Without prior attitude, star sensor first enters into initial attitude establishment mode when it starts working. By matching and identifying the measured full-sky star images captured by an imaging device, the precise initial attitude of star sensor is calculated and established. Then, star sensor enters into tracking mode.

2. Searching for Guide Star and Acquiring Star Information

In accordance with the attitude of star sensor, the current direction of boresight of the star sensor can be calculated. With the boresight pointing to the very direction, information on the guide stars within a certain range of a celestial area can be

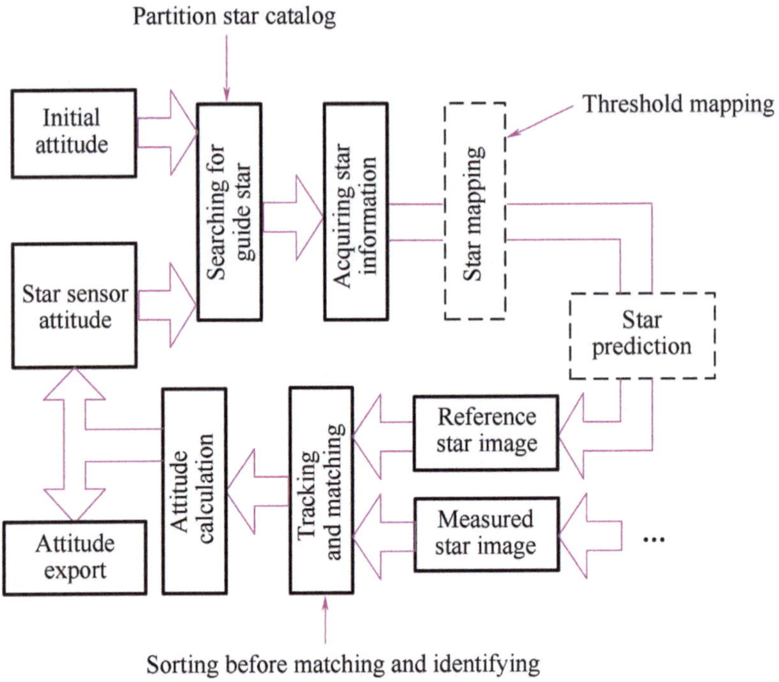

Fig. 6.3 Process of star tracking

obtained. Information of stars (including the index number, star magnitude, coordinates of the right ascension and declination and other information of stars) are found from the GSC. These stars are the ones that can be captured by image sensor under the current attitude. At this stage in the process of tracking, the partition star catalog is used. See Sect. 2.1 for details.

3. Star Mapping

Star images captured by an image sensor are based on image coordinate system (coordinates of stars are stored as pixels), while coordinates of guide stars stored in the star catalog are based on the celestial coordinate system. Hence, the coordinates of guide stars in the celestial area, which is expressed with respect to the celestial coordinate system, should be transformed and expressed in the image coordinate system (i.e., transforming the right ascension and declination information of guide star in star catalog into positional coordinate information on the image sensor). Threshold mapping is utilized at this stage in the process of tracking.

4. Star Prediction

When star sensor is operating, its angular velocity keeps changing. During tracking and matching, the movement of stars in the previous frames of known star images can be used to predict the position of stars in the next frame of star image. In this way, the success rate of matching identification can be significantly improved in tracking and matching. However, the introduction of star prediction increases the amount of calculation. Therefore, whether star predictions are to be introduced or not should depend on the circumstances.

5. Reference Star Image

In accordance with prior attitude information of star sensor, the reference star image is generated through procedures such as guide star searching, star mapping, star prediction and so on. Reference star image is not the actual star image captured by image sensor. Star information in the reference star image is already known.

6. Measured Star Image

Star images captured by image sensor and preprocessed are called measured star images. Star information in the measured star image is unknown.

7. Tracking and Matching

For a random star (marked by ★) in the reference star image, measured star (marked by ☆) is searched within its neighborhood *r* in the measured star image, as shown in Fig. 6.4. If no such measured star or more than one such star exists in this neighborhood, there is no correct match for the star in the measured star image. As Fig. 6.4 demonstrates, within the neighborhood of guide star No. 8, two measured stars, i.e. 7′ and 11′, are spotted. Hence, it is considered that there is no corresponding match for star No. 8. If there is one and only one measured star in the neighborhood, then this measured star in the measured star image can be successfully identified. In the process of tracking, the approach of sorting before matching is adopted.

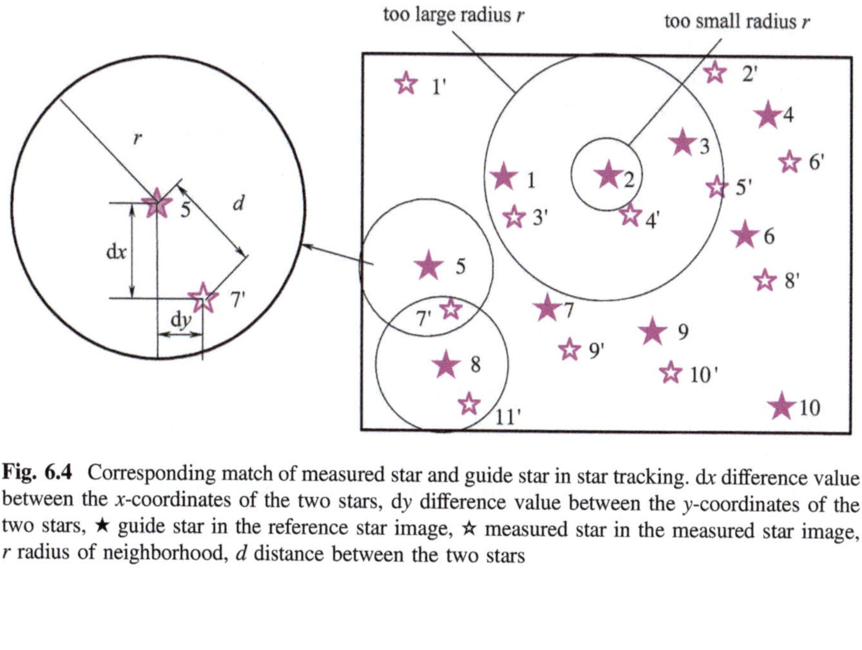

Fig. 6.4 Corresponding match of measured star and guide star in star tracking. d*x* difference value between the *x*-coordinates of the two stars, d*y* difference value between the *y*-coordinates of the two stars, ★ guide star in the reference star image, ☆ measured star in the measured star image, *r* radius of neighborhood, *d* distance between the two stars

The value of the neighborhood radius r is related to the angular velocity of star sensor. With too large a value, the number of stars that can be successfully matched and identified may decrease. For star No. 2 in Fig. 6.4, three measured stars, 3′, 4′ and 5′, will be spotted if the value of its neighborhood radius r is too large. The correct match for star No. 2, namely 4′, cannot be successfully matched and identified in this case. If the value of r is too small, however, 4′ will be out of the neighborhood, resulting in an unsuccessful match again.

8. Attitude Calculation

In accordance with the intrinsic parameters of star sensor, the attitude of star sensor with respect to the celestial coordinate system can be calculated by using the coordinates of a tracked measured star and the coordinates of its corresponding guide star in the star image.

On the basis of the calculated attitude of star sensor, the current pointing direction of boresight of star sensor can be computed. Information on guide stars within a certain range of the celestial area in this direction of boresight can be obtained. Making use of the imaging model of star sensor, the reference star image of the next frame can be obtained and used in the identification of the measured star image in the next frame. The tracking of measured stars in the measured star image is thus realized following these cyclic procedures.

It is clear from the illustration of star tracking that strategies such as partition star catalog, threshold mapping, sorting before matching and identification and others are adopted, accelerating the speed of star tracking. Among these strategies, partition star catalog divides the whole celestial area into several sub-celestial areas. In this way, only the sub-celestial areas adjacent to the pointing direction of the boresight of star sensor, instead of the whole celestial area, are searched in guide star mapping, reducing the number of guide stars to be retrieved and accelerating the speed of guide star indexing. In threshold mapping, with the demanded accuracy of attitude calculation, a threshold of the number of tracked stars is set so that guide star mapping is conducted only when the number of tracked stars is smaller than the threshold. Hence, the frequency of mapping, as well as the mean number of tracked stars, is reduced. Sorting before matching means that stars are arranged and ranked in accordance with their coordinate values in the star image before matching and identification, reducing meaningless matching of stars too far apart. In addition, with the introduction of star spot prediction, the position of a measured star in the next frame of star image is predicted on the basis of the tracking results of previous frames. Consequently, the neighborhood radius used is relatively small and the number of stars successfully tracked can be increased under the same circumstances.

For detailed discussion of the guide star indexing, threshold mapping and sorting before matching in star tracking algorithm, the identification results of star images in the kth and previous frames are defined as prior information and the task is to track and identify the current $(k + 1)$th frame of the measured star image. Two star images in Fig. 6.4 are regarded as the reference star image and measured star image of the $(k + 1)$th frame respectively.

6.2.2 *Guide Star Indexing by Using Partition of Star Catalog*

The partition of celestial areas is similar to the one introduced in Sect. 2.1. Guide star indexing by using partition of star catalog is conducted in the following way:

① Based on the results of previous tracking and identification, the direction vector of the boresight of star sensor is calculated;
② The subblock, whose medial axis vector is closest to the direction vector of the boresight most, is spotted in the partition star catalog;
③ This subblock and its adjacent subblocks (as shown in Fig. 6.5) constitute a sub-celestial area and the index number of guide stars in the area are stored;
④ In accordance with the stored index number of guide stars, corresponding guide stars are found in the star catalog and the positional information of guide stars are obtained. With the perspective projection transformation model, guide stars projected on the array plane of the imaging device of star sensor can be screened. In this way, the quick search of guide stars is accomplished.

With the partition star catalog, only nine subblocks, instead of the whole celestial area, are searched for the guide star. The use of partition star catalog narrows down the search area and accelerates the speed of searching and tracking.

6.2.3 *Threshold Mapping*

In the process of tracking, the generation of reference star, i.e. star mapping, is most time-consuming. It is quite a waste of time if each tracking has to undergo coordinate transformation. The aim of tracking is to calculate the attitude of star sensor by tracing the stars. To guarantee proper tracking, more than three stars should be

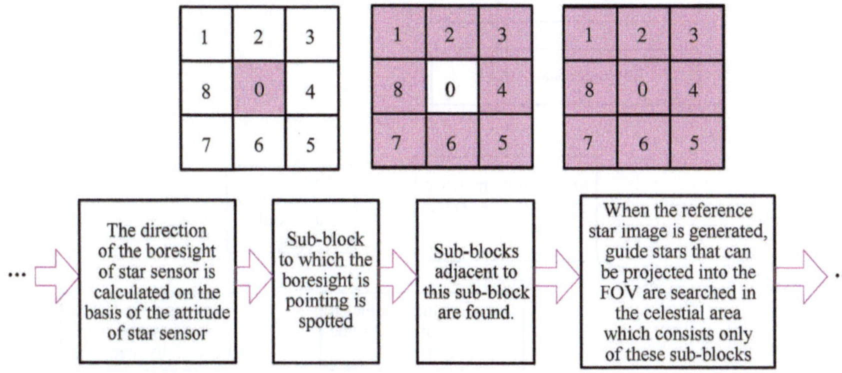

Fig. 6.5 Use of partition star catalog

tracked. Generally speaking, with the successful tracking of six to ten stars, the attitude of star sensor calculated can be accurate enough. Hence, threshold mapping is adopted in order to reduce the frequency of star mapping.

In threshold mapping, a star number threshold, defined as δ, is set. Star mapping is not conducted unless the number of measured stars successfully tracked is smaller than δ. Figure 6.6 presents the process of threshold mapping.

The detailed process is as follows:

① Firstly, a threshold δ is set. δ can be relatively large so that tracking algorithm can perform better in terms of reliability and attitude accuracy.
② Star mapping is not carried out unless less than δ stars have been matched and identified in the kth frame of the measured star image and its reference star image. When star mapping is not conducted, information of the measured stars in the measured star image (including right ascension, declination, magnitude, star index number and other information) is deemed to be the information of its matching guide stars in the reference star image. The matched and identified measured star image is directly considered as the $(k + 1)$th frame of the reference star image (only the information of successfully identified stars is kept). In this way, matching can be carried out simply between adjacent star images on the basis of prior information, with no need to generate a reference star image again.
③ If less than δ star are matched and identified in the kth frame of the measured star image, star mapping is conducted and the $(k + 1)$th frame of the reference star image is generated.
④ Similarly, the $(k + 1)$th frame of the reference star image and its measured star image are matched and identified. Star tracking is thus accomplished following these cyclic procedures.

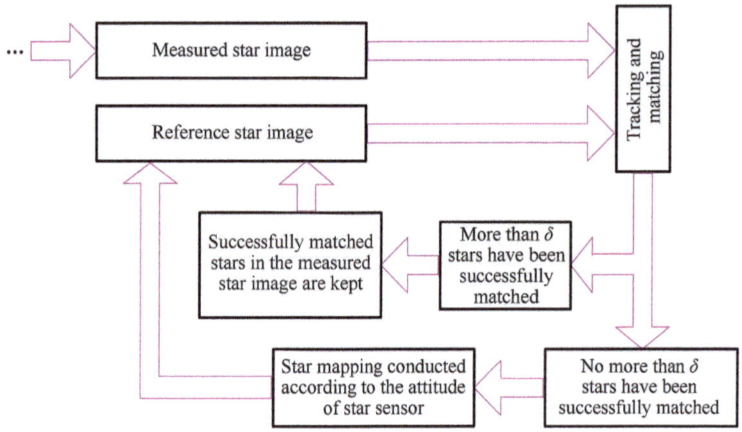

Fig. 6.6 Threshold mapping

6.2.4 Sorting Before Matching and Identification

When matching reference star images and measured star images, it is not necessary to compare those stars which are far apart. Thus, the strategy of sorting before matching and identification is utilized, i.e., star spots in the two star images are ranked in ascending order in accordance with their *x*-coordinates. (Fig. 6.7 demonstrates the ranking of stars in the two star images in Fig. 6.4.) Then, star matching and identification are carried out.

(a) Sorting of the $(k + 1)$th frame of reference star image
(b) Sorting of the $(k + 1)$th frame of measured star image

For better illustration, the sorted $(k + 1)$th frame of the reference star image and its measured star image are marked as sequence A and sequence B, respectively.

The matching and identification of the two star images shown in Fig. 6.4 can be regarded as the matching process of the two sequences. Take the matching of star No. 2 in A and stars in B as an example (as shown in Fig. 6.8), the detailed process is as follows:

① In order to reduce computational work and accelerate calculation speed, the difference between d*x* (*x*-coordinate difference) and d*y* (*y*-coordinate difference), instead of the distance between two stars *d*, is compared. The equations are as follows:

$$dx = |x - x'|, \quad dy = |y - y'| \tag{6.1}$$

② The comparison of star No. 2 in A sequence and stars in B sequence starts with 3′. 3′ is the first star in B-sequence whose d*x* becomes smaller than *r* when star

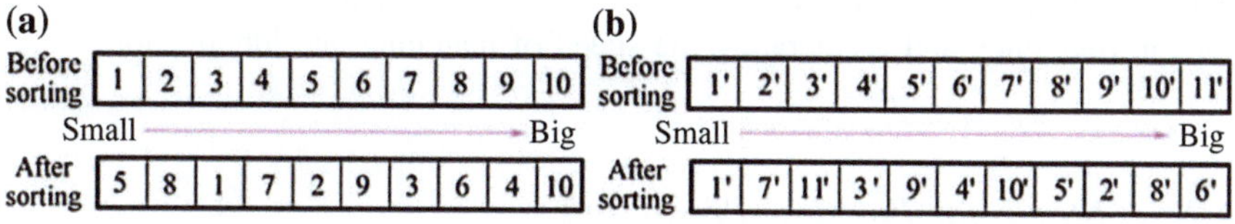

Fig. 6.7 Sorting of two star images

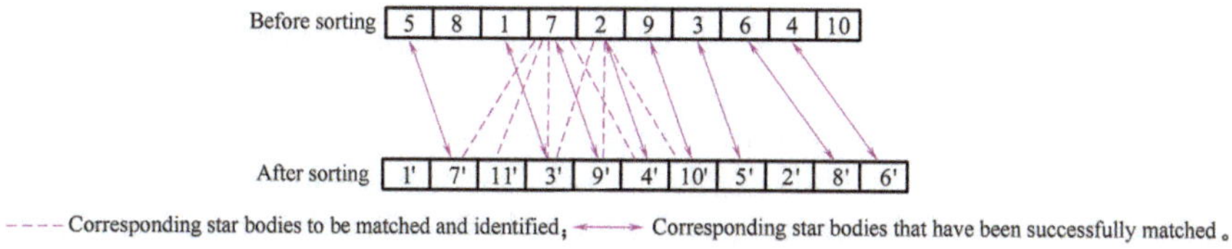

Fig. 6.8 Sorting before matching and identification

No. 7 in A and those in B are compared. In other words, the dx values of star No. 7, as well as 7′ and 11′ before 3′, are all larger than r. Since the stars are arranged in ascending order, the dx values of star No. 2 after No. 7, and stars No. 7′ and 11′ should be larger than r. Hence, there is no need to compare star No. 2 with them.

③ In the comparison, if the dx values of star No. 2 and B-sequence stars are smaller than r, then this No. 2 star is compared with the next star in B sequence. In this case, No. 2 is to be compared successively with stars No. 3′, 9′, 4′ and 10′.

④ When star No. 2 is compared with 10′, it is found that their dx value is larger than r. The comparison of star No. 2 with B-sequence stars should be terminated at this time as there is no need to continue the comparison. Since the stars are arranged in ascending order, the dx values of star No. 2 and stars No. 5′, 2′, 8′ and 6′ (which are after star No. 10′), are definitely larger than r. There is no need to make further comparison.

⑤ During the comparison of star No. 2 and B-sequence stars, only the dx and dy values of No. 2 and 4′ are smaller than r simultaneously, signifying that there is only one star in the neighborhood of star No. 2. Thus, 4′ is matched and identified. The guide star information of 4′ can be obtained from the identification results of the previous frame of star No. 2.

⑥ Similarly, stars No. 9, 3, 6, 4 and 10 are compared with B-sequence stars, accomplishing the matching and identification of this frame of the measured star image.

With the approach of sorting before matching and identification, unnecessary comparisons are effectively eliminated. As demonstrated in Fig. 6.8, star No. 2 in the $(k + 1)$th frame of the reference star image is simply to be compared with stars No. 3′, 9′, 4′ and 10′ in the measured star image. It is not necessary to compare star No. 2 with other stars. With sorting before matching and identification, only the stars with approximate coordinate values are to be compared, reducing the comparison frequency and accelerating the speed of matching and identification.

6.2.5 Star Spot Position Prediction

When star sensor is operating, the attitude angular velocity of its carrier keeps changing. The range of attitude angular velocity is defined as 1°–5°/s. Requirements for the value of neighborhood radius r varies significantly at different angular velocities.

If the neighborhood radius r used in tracking and identification is a constant, this value may be suitable at one velocity but inappropriate at another, either larger or smaller. As a result, the tracking may be low in efficiency, or cannot be accomplished in some cases.

The value of the neighborhood radius has a direct impact on the effects of tracking and matching, as is shown in Fig. 6.4:

① When the value of the neighborhood radius r is small (small circle), there is only one star, 4′, in the neighborhood of star No. 2. Hence, the two stars are successfully matched.

② When the value of the neighborhood radius r is large (big circle), there are two stars, 3′ and 4′, in the neighborhood of star No. 2. Hence, 4′ cannot be tracked and identified.

Due to the big differences in the neighborhood radius demanded by different angular velocities, it is inappropriate to set a single constant value for the neighborhood radius.

Star spot position prediction can be used to solve the problem. There are two approaches which can be taken in order to predict the position of star spots. The second approach is utilized in this book.

① The position of stars in the FOV is predicted through the accurate estimation of attitude. As a widely used method, it estimates current attitude information on the basis of precise information of attitude and angular distance. Then, the positional coordinates of stars in the FOV are calculated in accordance with the current estimated value of the attitude. The strength of this method is that the estimated position acquired through this approach is usually accurate. However, the calculation is relatively complex and requires the use of a very complicated filtering algorithm.

② The imaging position of star spots in the FOV is estimated by using the image and angular velocity. In accordance with the positional changes of stars tracked in the previous moment, the position of the star is predicted in the current FOV. This strategy is easy to accomplish and fast to calculate [8]. It is noteworthy that when full-sky star identification mode is transformed into tracking mode, the star spot position at the initial moment cannot be estimated since there is no previous record to be used for tracking. A larger neighborhood radius has to be used for tracking. After one successful tracking, tracking information of the previous frames can be utilized to predict the position of the tracked star in the next frame. The neighborhood radius used at this time can be reduced in its value.

With the adoption of star spot prediction, the rate of matching and identification is significantly improved, which also increases the number of stars to be tracked and matched to some extent.

6.3 Simulations and Results Analysis

Simulation experiments are conducted to evaluate the comprehensive performance of the rapid star tracking algorithm. The contents of simulation experiments include selection of star tracking parameters, influence of star position noise and attitude angular velocity on star tracking, and time of tracking and processing.

6.3.1 *Selecting Star Tracking Parameters*

1. Influence of Threshold Value on Tracking Velocity

It is clear in Fig. 6.9 that eight measured stars are correctly matched and identified in the measured star image.

If the threshold δ is defined as 10, then the number of tracked stars is smaller than the threshold. Star sensor's attitude is calculated and star spot mapping is accomplished, generating a reference star image, as shown in Fig. 6.9a.

If the threshold δ is defined as 6, then the number of tracked stars is bigger than the threshold and star spot mapping is not conducted. The previous frame of the measured star image is considered as the reference star image for the next frame, as shown in Fig. 6.9b.

Star spot mapping is a crucial and time-consuming process in star tracking. With a properly set threshold, the frequency of star mapping and the mean number of tracked stars are effectively reduced. The tracking speed is also improved by shortening the time for tracking, matching and identifying the reference star image and the measured star image.

Figure 6.10 demonstrates a comparison of tracking velocities with different threshold values (initial attitude: yaw angle 5°, pitch angle 60° and roll angle 10°; final attitude: yaw angle 15°, pitch angle 70° and roll angle 20°). The upper curve is

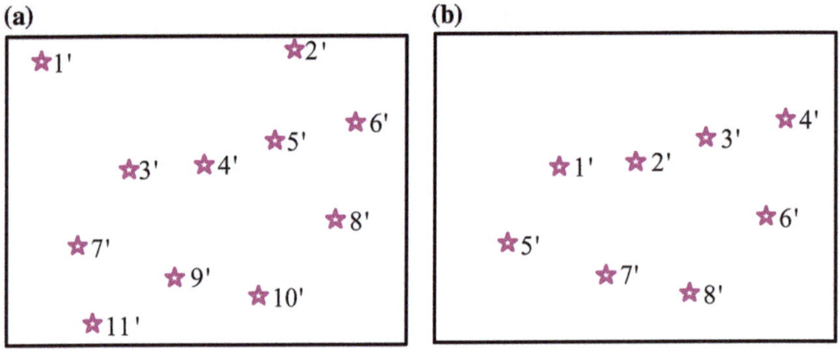

Fig. 6.9 Influence of δ value on the generation of a reference star image

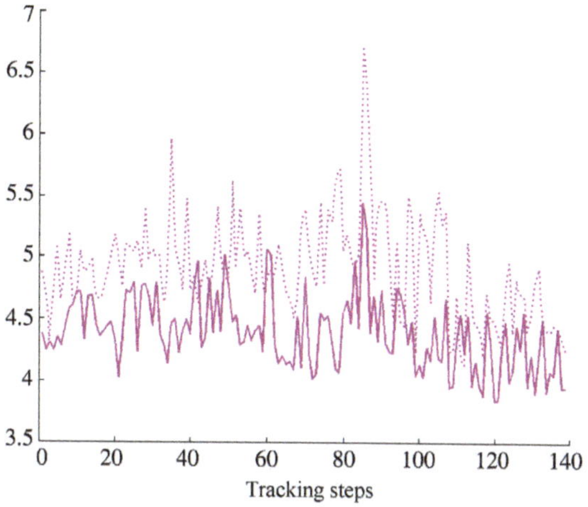

Fig. 6.10 Threshold value's influence on star tracking velocity

the tracking time with a threshold of 30, while the lower one demonstrates the tracking time with a threshold of 6.

It is clear from the illustration that as the threshold increases, a longer time is required for each tracking step. Therefore, given that the attitude measurement remains accurate, the value of the threshold should be kept as small as possible.

2. Threshold Value's Influence on the Success Rate of Star Tracking

To speed up star tracking, the threshold value should be kept as small as possible in theory. However, the value should not be too small. The reasons lie in the following aspects:

① At least three stars are required to be successfully tracked for the calculation of attitude. Hence, the threshold value should not be smaller than 3.
② Occasionally, stars successfully tracked in the current frame of measured star image may move out of the FOV in the next frame, resulting in tracking failure. It often occurs when the angular velocity is high and the displacement of a measured star in adjacent frames of images is great.

When the threshold is set to be 6 and eight stars have been successfully tracked, no star spot mapping is required in this kind of star tracking. However, if six of these tracked stars move out of the FOV in the next frame of the measured star image, only two stars can be successfully tracked at most. Under this circumstance, star tracking may fail and attitude calculation cannot be carried out.

As Fig. 6.11 shows, when the threshold is 4, the success rate of tracking, which is around 55%, is relatively low. With the threshold bigger than 8, the success rate of star tracking is relatively high, remaining above 95%.

Since the threshold value should be kept as small as possible, the threshold can be set as 8 in star tracking. In this way, the success rate can be ensured and the processing time can be effectively shortened in star tracking.

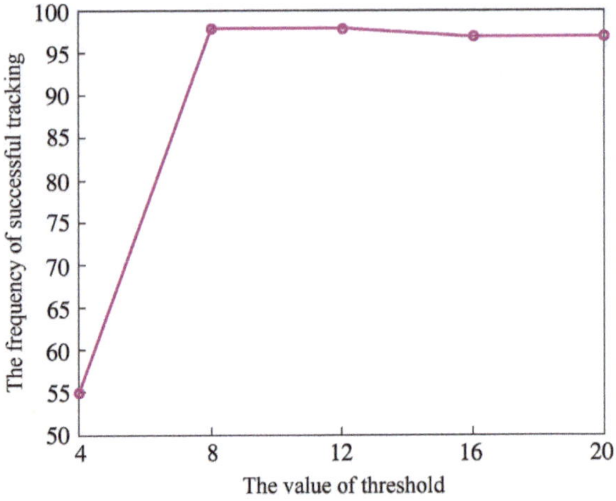

Fig. 6.11 Threshold value's influence on the success rate of star tracking

3. Neighborhood Radius' Influence on the Success Rate of Star Tracking

Figure 6.1 illustrates the minimum values of the neighborhood radius with and without star spot prediction.

As is shown in Table 6.1, with different angular velocities of attitude motion, the minimum values that satisfy the neighborhood radius r in normal tracking and identification vary significantly. The value of the neighborhood radius r should grow correspondingly as the angular velocity increases. It is also clear that a smaller value of neighborhood radius can be adopted with the introduction of star spot prediction in star tracking.

The value of the neighborhood radius also influences star tracking. If the radius is too large, more stars will appear in the FOV for matching and identification during the process of star tracking. The corresponding measured star cannot be correctly identified, decreasing the number of measured stars in the measured star image and guide stars in the reference star image that may be successfully matched and identified. On the contrary, if the radius is too small, the corresponding measured star may fall out of the neighborhood radius.

Figure 6.12 reflects the changes in the success rate of star tracking with the increase in the value of the neighborhood radius and the introduction of star spot prediction.

Table 6.1 Minimum values of neighborhood radius under different circumstances

Angular velocity of attitude motion (°/s)	Neighborhood radius without star spot prediction (pixel)	Neighborhood radius with star spot prediction (pixel)
1	8	4
2	15	4
3	21	5
4	29	5
5	35	7

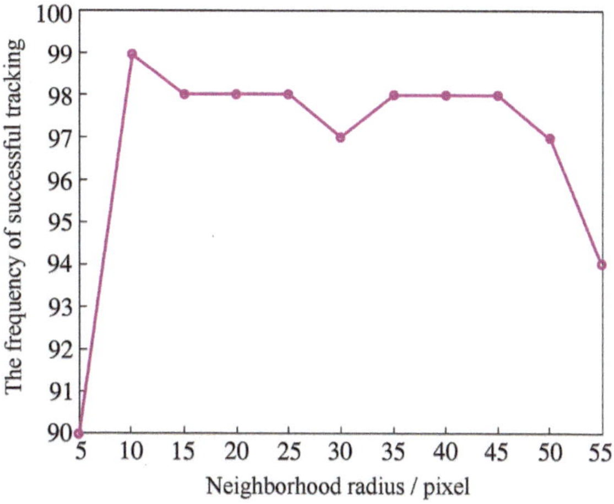

Fig. 6.12 Neighborhood radius' influence on the success rate of star tracking

As shown in Fig. 6.12, if the neighborhood radius is too large or too small, the success rate of tracking will decrease. The highest success rate of star tracking can be acquired when the neighborhood radius is about 10.

6.3.2 Influence of Star Position Noise on Star Tracking

In order to study the influence of star position noise on star tracking, position noise is introduced into the generated measured star image in simulation experiments. Following Gaussian distribution, the position noise has a mean value of 0 and standard deviation of 0–2 pixels.

1. Influence of Position Noise on the Success Rate of Star Tracking

In simulations, star tracking that can accomplish the predetermined tracking process is considered to be successful. In other words, successful star tracking refers to those that can track and identify measured star images generated during each step with the attitude calculated on the basis of the results of tracking and identification within the range of error tolerance. Once star tracking fails to track one frame and cannot calculate the attitude, it is deemed as a failure.

Gauss noise, with a mean value of 0 and standard deviation of 0 (i.e., no noise), 0.5, 1.0, 1.5 and 2, respectively, is introduced into the star positions in the measured star image. A hundred tracking processes are randomly selected to run the test.

Figure 6.13 demonstrates the results of tracking simulation. It is found that though the success rate of tracking decreases as the standard deviation of position noise increases, it remains above 95%.

In simulation, star tracking fails mostly because the number of stars that are successfully tracked is too small. This number cannot meet the minimum demand for attitude calculation. As a result, the attitude of star sensor cannot be correctly

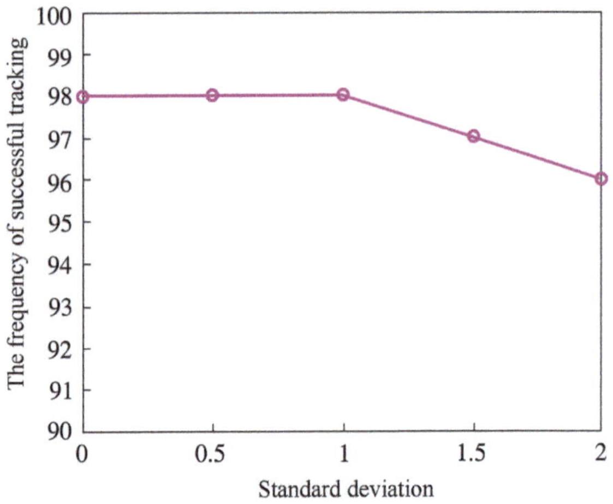

Fig. 6.13 Influence of position noise on the success rate of star tracking

calculated and the next frame of reference star image cannot be generated. Thus, the tracking process is terminated and star tracking fails.

2. Influence of Position Noise on the Export Accuracy of Attitude in Star Tracking

As can be seen from the attitude calculation equation, the computation process is related to the coordinates of star position. The accuracy of the star position in the measured star image has a direct bearing on the measurement precision of the starlight vector in any star sensor coordinate system. The attitude export accuracy during star tracking is further affected by this precision.

Gauss noise, with a mean value of 0 and standard deviation of 0–2, is introduced into the star positions in the measured star image. Several tracking processes are randomly selected to run the test.

A random attitude of star sensor is selected. Its initial attitude has a yaw angle of 300°, a pitch angle of 40° and a roll angle of 0°. The final attitude has a yaw angle of 310°, a pitch angle of 50° and a roll angle of 10°. Figure 6.14 illustrates the influence of position noise on the accuracy of the star sensor attitude. The curve with relatively small fluctuations represents the difference value between the calculated attitude value and the true value when the standard deviation of position noise is 0.5. The curve with relatively large fluctuations reflects the difference value between calculated attitude value and the true value when the standard deviation of position noise is 2.

It is clear from Fig. 6.14 that the increase in position noise reduces attitude accuracy. Therefore, star centroiding of star sensor should be maintained as accurately as possible in actual use in order to improve the precision of star sensor's attitude measurement.

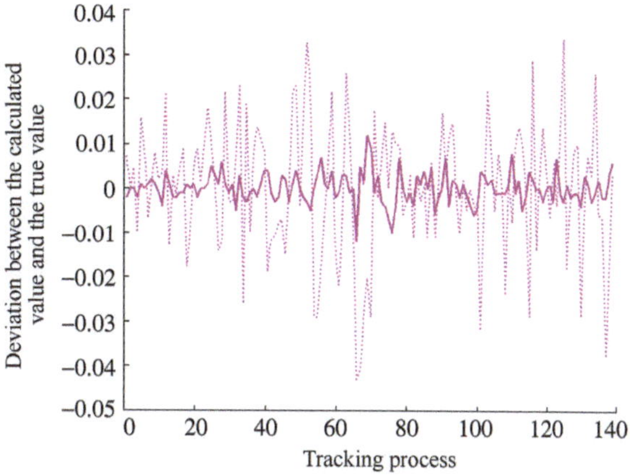

Fig. 6.14 Influence of position noise on attitude accuracy

6.3.3 Influence of Star Sensor's Attitude Motion on Star Tracking

In simulation experiments, the measured star images in each tracking step are simulated on the basis of the given attitude of star sensor. With the set initial and final attitudes of star sensor, the given attitude of star sensor changes in the given manner.

In actual use, the attitude of star sensor does not necessarily change in only one manner. The impact of different manners on attitude calculation is studied by imposing various changes on the given attitude. In this way, the simulation experiments of star tracking can be more practical and the tracking algorithm can perform better.

A random attitude of star sensor is selected. Its initial attitude has a yaw angle of 190°, a pitch angle of −70° and a roll angle of 0°. The final attitude has a yaw angle of 200°, a pitch angle of −60° and a roll angle of 10°. With this given attitude, situations in which the attitude is subject to linear variation and conic variation are analyzed respectively.

1. Linear Variation of the Given Attitude

The given attitude changes in accordance with equation

$$Y = AX + C$$

Here, Y stands for the attitude in each tracking step, A for the coefficient of the linear equation, X for the steps of tracking, and C for the initial attitude of tracking. The given attitude in each tracking step changes linearly as the tracking process differs.

Figure 6.15 demonstrates the comparison between the given value and the calculated value of yaw angle in each tracking step when the given attitude changes linearly. On the right of the Figure is an enlarged illustration of part of the tracking

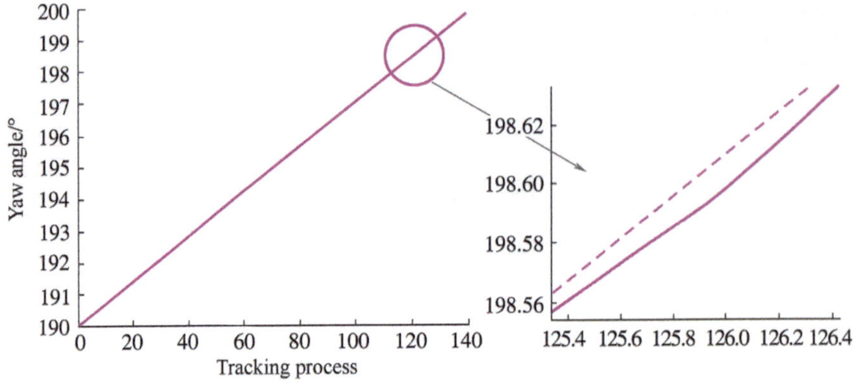

Fig. 6.15 Tracking curve of the linear variation of the given attitude

curve. The dotted line represents the value of the given attitude, while the solid line stands for the calculated value of the attitude. It is clear from the Figure that the curve of the calculated value is basically in line with the curve of the given value.

2. Conic Variation of the Given Attitude

The given attitude changes in accordance with equation

$$Y = AX^2 + C$$

Here, Y stands for the given attitude in each tracking step, A for the coefficient of the conic equation, X for the steps of tracking, and C for the initial attitude of tracking. With this equation, the given attitude in each tracking step changes in accordance with the conic curve.

Figure 6.16 demonstrates the comparison between the given value and the calculated value of yaw angle in each tracking step when the given attitude changes in

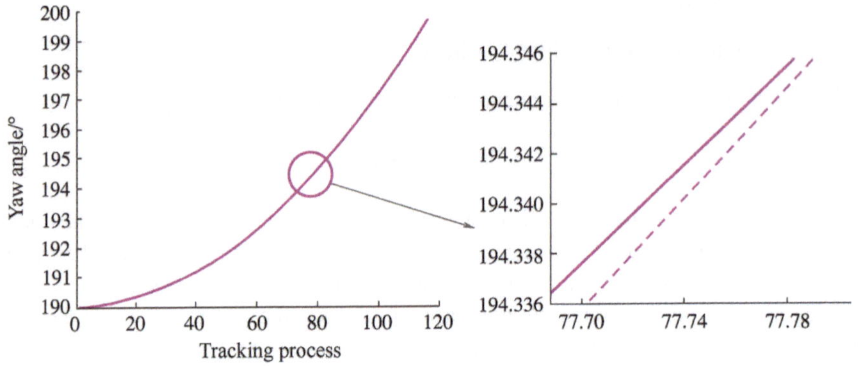

Fig. 6.16 Tracking curve of the conic variation of the given attitude

accordance with the conic curve. On the right of the Figure is an enlarged illustration of part of the tracking curve. The dotted line represents the value of the given attitude, while the solid line stands for the calculated value of the attitude. It is clear from the Figure that the curve of the calculated value is basically in line with the curve of the given value.

6.3.4 Speed of Star Tracking

In order to assess the rapidity of the tracking algorithm, simulation tests on the processing time of star tracking are carried out. When the threshold is set to 8, i.e., the number of stars being tracked in the FOV is smaller than 8, star mapping is conducted.

The initial attitude of the simulation is set to have a yaw angle of 190.707°, a pitch angle of −88.168° and a roll angle of 1.210°. The final attitude has a yaw angle of 200.707°, a pitch angle of −78.168° and a roll angle of 11.210°. Figure 6.17 presents the statistical result of the time spent on 140 frames of star tracking. The dotted line stands for the tracking time spent before improvement, and the solid line for the tracking time spent after improvement (by using zone catalog-based quick retrieval of guide stars, threshold mapping, sorting before tracking and other strategies). The statistical results demonstrate that an average of 12 ms is taken in each tracking step before the improvement and 6 ms after the improvement.

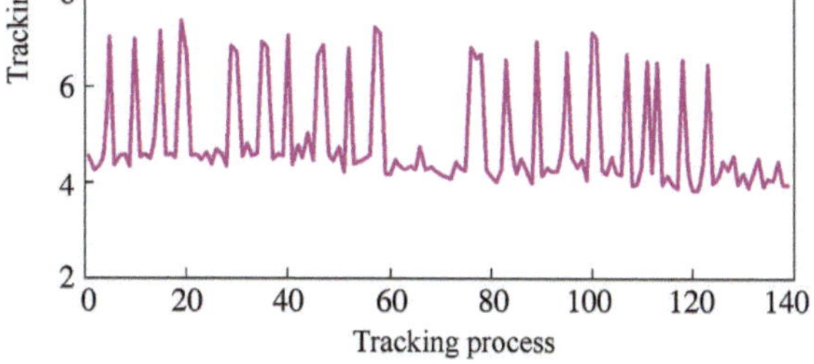

Fig. 6.17 Statistical results of processing velocity of star tracking

References

1. Jiang J, Zhang GJ, Wei XG et al (2009) Rapid star tracking algorithm for star sensor. IEEE Aerosp Electron Syst Mag 23–33
2. Jiang J, Li X, Zhang G, Wei X (2006) Fast star tracking technology in star sensor. J Beijing Univ Aeronaut Astronaut 32(8):877–880
3. Jiang J, Li X, Zhang G, Wei X (2006) A fast star tracking method in star sensor. J Astronaut 27(5):952–955
4. Li X (2006) Fast star tracking technology in star sensor. Master's thesis of Beijing University of Aeronautics and Astronautics, Beijing
5. Laher R (2000) Attitude control system and star tracker performance of the wide-field infrared explorer spacecraft. American Astronomical Society, AAS 00-145:723–751
6. Wang G et al (2004) Kalman filtering algorithm improvement and computer simulation in autonomous navigation for satellite. Comput Simul 27(1):33–35
7. Yadid PO et al (1997) CMOS active pixel sensor star tracker with regional electronic shutter. IEEE Trans Solid-State Circ 32(3):285–288
8. Samaan MA, Mortari D, Junkins JL (2001) Recursive mode star identification algorithms. Space flight mechanics meeting, Santa Barbara, CA, AAS 01-194, pp 11–14

Chapter 7
Hardware Implementation and Performance Test of Star Identification

As an aerospace product, star sensor must meet the demand for miniaturization. Currently, star sensor uses embedded design schemes and its integration level is increasingly high so as to minimize its weight, power consumption, and size. The star identification algorithm generally runs on embedded processors. Meanwhile, to store GSC and navigation feature databases, it is also necessary to extend the peripheral memory. RISC (Reduced Instruction Set Computer) processor based on ARM (Advanced RISC Machines) has such characteristics and advantages as low power consumption, low cost, and good performance. It is used widely by numerous star sensor products as a core device of the data processing unit of star sensor.

After simulation experiments using simulated star images, the star identification algorithm needs to be further tested to investigate its performance when it is closer to the on-orbit operational status of star sensor. Generally, there are two ways of testing: one that uses a star field simulator to conduct hardware-in-the-loop simulation and verification and one that conducts the field test of star observation. The former can be done in a laboratory and is not restricted by weather conditions or geographical positions. Through flexible configurations of simulation star images, diversified function tests and verifications can be done. The latter can obtain the actual star images that are closest to the operational status of star sensor, but it is subject to the influences of atmospheric environment, climatic conditions, and geographical positions.

This chapter introduces the hardware implementation process of the star identification algorithm by taking the RISC processor as an example, and describes its two testing methods, i.e., hardware-in-the-loop simulation and verification and field test of star observation.

G. Zhang, *Star Identification*, DOI 10.1007/978-3-662-53783-1_7

7.1 Implementation of Star Identification on RISC CPU

The circuit system of star sensor is generally put into two parts: front end and back end. The former mainly fulfills the driving of the image sensor and low-level processing of star image. The centroid coordinates of star spot obtained through front end processing are often implemented by FPGA or CPLD. The latter mainly fulfills star identification, star tracking, attitude establishment, etc. The final output of attitude information is often implemented by RISC or DSP processor. This section mainly introduces the structure of RISC processing circuit at the back end of star sensor and the implementation of star identification algorithm [1, 2].

7.1.1 Overall Structural Design of RISC Data Processing Circuit

The RISC processor enjoys better pipeline function and requires fewer gate circuits for its implementation. Compared with other microprocessors, it is lower in power consumption and cheaper. Thus, many star sensors choose RISC processor as their processor. For example, SETIS star sensor from German company Jena-Optronik uses advanced ASIC chip technology with 16-bit RISC controller (PMS610) at its core, reducing the cost by 50%. American JPL (Jet Propulsion Laboratory) takes the lead in designing a micro star sensor that uses CMOS image sensor and 32-bit RISC processor. Its power consumption is just 500 mW when the voltage is 5 V.

With RISC processor at its core, RISC data processing circuit adds communication interface module, memory module, JTAG interface module, RS232 serial communication module, and power module at the periphery to implement such functions as debugging, computation, and communication. Figure 7.1 shows the framework of the RISC data processing circuit.

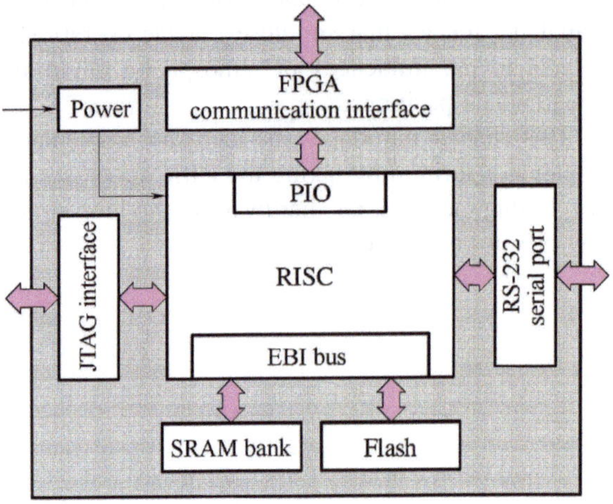

Fig. 7.1 Framework of RISC data processing circuit

Some memory space must be allocated in advance for the normal running of star identification algorithm. The demand of star identification algorithm for memory space is in two aspects:

- The running code segment, global variable, and stack of the program itself require some memory space.
- Star identification algorithm must depend on guide database. Generally, GSC and navigation feature database that form guide database require relatively large memory space.

Due to the small internal memory capacity of ARM RISC processor, it is thus indispensable to extend external memory in the circuit design. By adding SRAM (Static RAM) memory bank to EBI bus of RISC processor, Flash memory can implement the expansion of memory space. Modified triangle algorithm based on angular distance matching can be taken as example. In designing software, it is preliminarily estimated that the program running itself requires about 1.6 MB of space, while guide database requires about 1.1 MB of memory space. Thus, the capacity of SRAM and Flash memory used should reach 2 MB.

Since the star image data processed by RISC data processing circuit comes from the FPGA circuit of the front end, the interface that communicates with FPGA must be included, which is implemented by the PIO (Parallel Input/Output) interface of RISC processor. In addition, the RS-232 serial port is also configured for attitude output.

At the development stage, the program used for implementing star identification and attitude establishment must be downloaded to the RISC processor for debugging. Thus, a JTAG interface is added to the RISC processing circuit to debug software by connecting with the simulator.

7.1.2 Selection of Primary Electronic Components

The primary electronic components of RISC data processing circuit include RISC processor and peripheral memory.

(1) RISC processor-AT91R40008

AT91R40008 is one of the AT91 ARM microprocessor series products of ATMEL Company [3]. Its kernel is ARM7TDMI processor, the structure of which is shown in Fig. 7.2. This processor is of 32-bit RISC structure with high performance and density and 16-bit instruction set. Besides, its power consumption is very low. Inside the processor, there are SRAM of 256 KB and numerous registers that can rapidly deal with unusual interruptions, thus making it suitable for occasions with real-time requirements. AT91R40008 can be directly connected to external memory. It has a fully programmable 16-bit external bus interface and can chip select eight peripherals, which speeds up access to memory and reduces the price of the

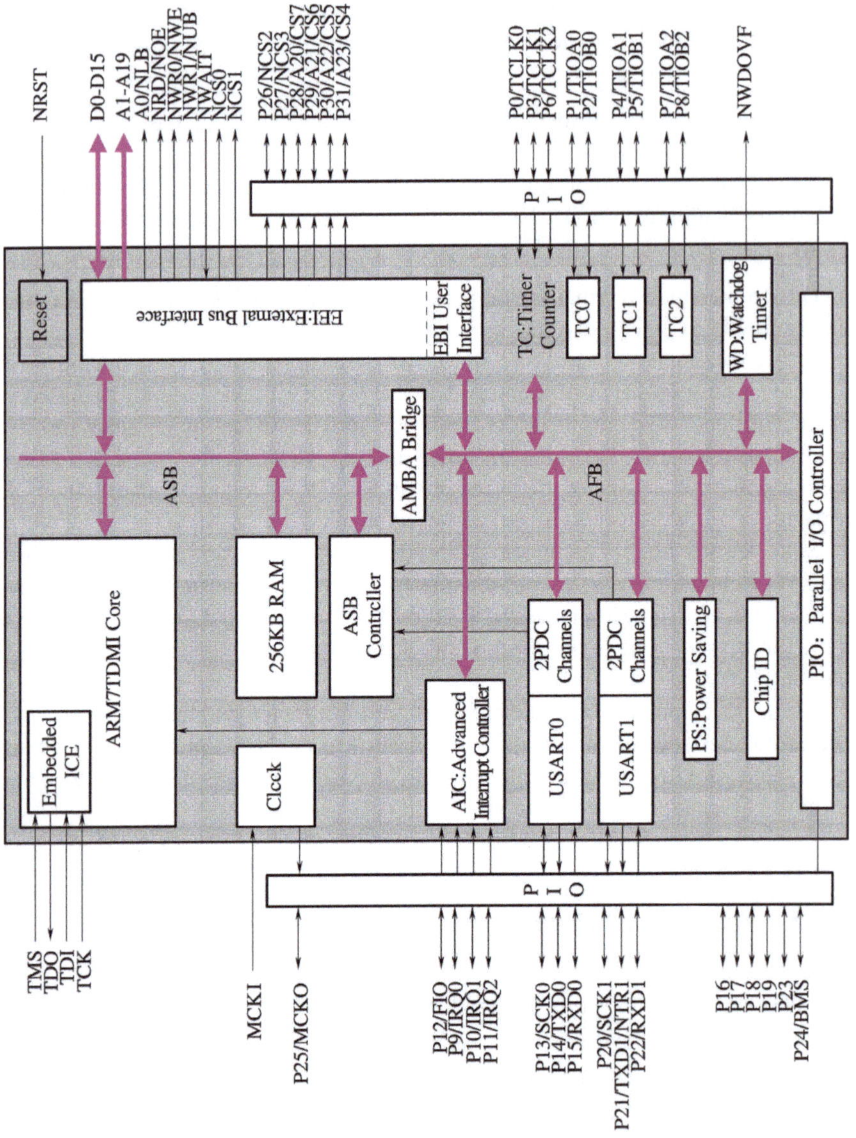

Fig. 7.2 Internal structure of AT91R40008

system at the same time. AT91R40008 supports 8 bit/16 bit/32 bit read and write (with SRAM of 256 KB inside) and interrupt control with eight priority grades. It has 32 programmable I/O interfaces and three 16 bit timers/counters supporting three external clocks input. AT91R40008 has two USART (Universal Synchronous/Asynchronous Receiver/Transmitter) units which can provide the function of full-duplex serial communication. AT91R40008 needs two levels for its power: VDDIO of 3.3 V and VDDCORE of 1.8 V. The maximum processing frequency of AT91R40008 can reach 60 MHz. But to ensure the steady operation of the system, its operation frequency is slightly reduced and crystal oscillator of 50 MHz is used.

(2) Peripheral Memory

Peripheral memory is connected to an external bus interface, mainly including SRAM and Flash.

SRAM includes fast read speed capability and programs mainly run here after start-up, which can improve the operating speed of star sensor. What SRAM selects is ISSI company's 16 bit IS61LV51216, which is of 512 KB, high speed and low level. Its read speed is 8 ns, and its power and ground pin are in the middle of the chip, reducing interferences. The chip select and read/write enable pins are $\overline{CE}$, $\overline{OE}$, $\overline{WE}$, respectively, which are all active low. Flash is mainly used for storing data and code. Even when it is powered off, the information will not be lost. What Flash selects is ATMEL company's AT49BV162A. Its read speed is 70 ns, the time of rapid word programming is 20 μs and the time of rapid sector erasure is 300 ms. Through $\overline{BYTE}$ pin, AT49BV162A can choose Byte or Word model. Since the external data bus interface of AT91R40008 is 16 bit, AT49BV162A chooses Word model. The VDD level of AT49BV162A is 3.3 V. With its VPP pin, the programming and erasure speed of Flash can be improved when the level is 5 V or 12 V. Being active low, the enabled pins of chip select and read and write are $\overline{CE}$, $\overline{OE}$, $\overline{WE}$, respectively.

7.1.3 Hardware Design of RISC Data Processing Circuit

(1) RISC Processor and Memory and Bus Interface Circuit

RISC processor and memory and bus interface circuit are connected mainly by address lines, data lines, and control pins such as chip select, enable, etc. The circuit is shown in Fig. 7.3.

There are both 20 address lines and 16 data lines between the RISC processor and SRAM and Flash, each being able to reach addressing of 2 MB. It is noticeable that ARM7 series of ARM processor have two processing statuses: thumb status with 16-bit instruction set and ARM status with 32-bit instruction set. In addressing, the last bit (A0) of the address is zero all along. Thus, when the address lines are connected between RISC processor, SRAM and Flash, the A1 bit of RISC processor is connected to the A0 bit of SRAM and Flash, forming an addressing space of 2 MB.

RISC processor connects chip select 1, i.e., $\overline{CS1}$, (NCS1 in Fig. 7.3) to the enable pins of SRAM, and allocates base address 0x02000000 to SRAM. Meanwhile, RISC processor connects chip select $\overline{CS0}$ (NCS0 in Fig. 7.3) to the

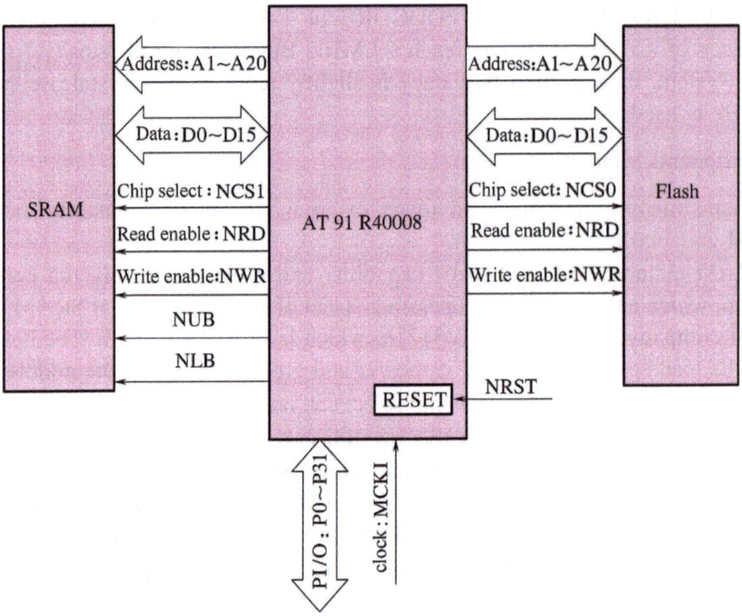

Fig. 7.3 RSIC processor, memory and bus interface circuit

enable pins of Flash, and allocates base address 0x01000000 to Flash. The read-write enable pins of RISC processor and memory can be connected accordingly. It should be noted that SRAM needs to be connected to the high 8-bit gating signal NUB and the low 8-bit gating signal NLB as well.

(2) SRAM Memory Circuit

SRAM memory circuit combines two 16-bit SRAM chips of 1 MB, extending the space to 2 MB. The SRAM memory circuit is shown in Fig. 7.4.

As shown in Fig. 7.4, there are two SRAMs: Bank0 and Bank1, each with a 16-bit data line. The two SRAMs select CS_BANK0 and CS_BANK1 as their chip select signals, respectively, which can be obtained through 74LVC139 decoder. The two chip select signals are related to address A20. When A20 is zero, CS_BANK0 is zero and CS_BANK1 is one. Here, SRAM0 is selected. When A20 is one, CS_BANK0 is one, and CS_BANK1 is zero. Here, SRAM1 is selected. Thus, a SRAM module of 2 MB is formed, extending the memory space.

The two SRAMs share the read enable signal $\overline{RD}$, the write enable signal $\overline{WE}$ (NRD and NWE in Fig. 7.4, respectively) and the high 8-bit gating signal NUB and the low 8-bit gating signal NLB.

(3) Interface Circuit Design of RISC Processor

When the RISC processor is at work, front end FPGA is required to provide the coordinates of star spots in the captured star images. RISC processor receives data

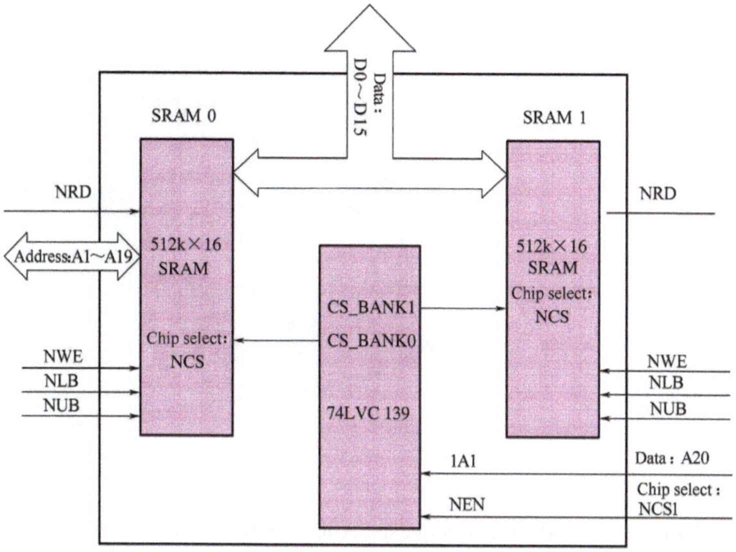

Fig. 7.4 SRAM memory circuit

from FPGA through PIO interface and then conducts star identification and attitude establishment. Data transmission shares 19 PIO interfaces, three of which are used for communication control, i.e., Status, Req, and Ack. The other 16 interfaces are used for data transmission.

To transmit the centroid data of star spots to RISC processor at the back end of star sensor, data transmission protocol between FPGA and RISC is defined. Figure 7.5 shows the time sequence of data transmission protocol.

Status: Operation status of FPGA, input signal;
Req: Read requests of RISC processor, initial value set to high, output signal;
Ack: Answering signal of FPGA, active high, initial value set to low, input signal;
Data: 16-bit data read by RISC processor, input signal.

Before communication, Req is high, Ack is low, and the interface state machine inside FPGA is idle. That Status is set to low means that a new frame of centroids data has been prepared to conduct data transmission. Once the communication starts, if the RISC processor reads the falling edge of Status, it can then set Req to

Fig. 7.5 Time sequence diagram of RISC processor and FPGA data transmission protocol

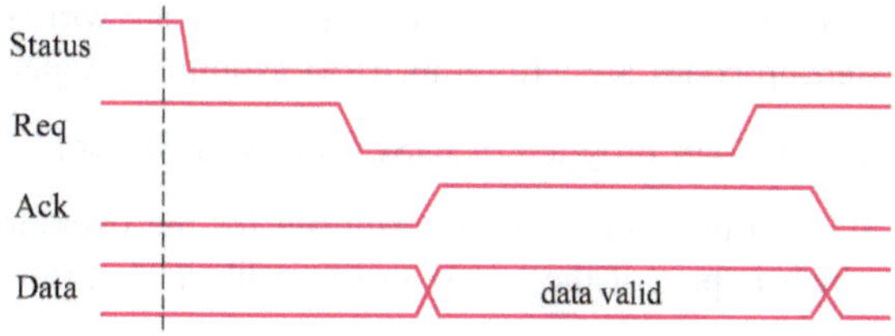

Fig. 7.6 Physical graph of RISC data processing circuit

low level and send out data read requests. After reading the low-level signal of Req, the state machine that implements interface protocol inside FPGA puts corresponding data to data lines and sets Ack to high at the same time. Here, the interface state machine inside FPGA is in the state of being able to read. Noticing that Ack ascends, the RISC processor reads the data, sets Req to low and then informs FPGA that the data has been read. After FPGA reads that Req is set to high, the state machine returns to the idle state and sets Ack to low. Thus, data transmission is finished. The centroid data of all the star spots in a frame of image are transmitted cyclically in this way.

It has been tested that the transmission time of the centroid data of each star spot is approximately 0.5 ms. Take the centroid data of 20 star spots in a frame of star image for example. The transmission time of the centroid data of each frame of star image is approximately 10 ms, which can meet the real-time demand.

Figure 7.6 shows the physical graph of RISC data processing circuit.

7.1.4 Software Design of RISC Data Processing Circuit

The programs that run on a RISC processor mainly include star identification and tracking programs. There are also some auxiliary programs such as start-up program, serial communication program, time testing program, peripheral data communication program, etc. The major codes are written in C language and part of the start-up codes are written in ARM assembly language.

(1) Start-Up Program of RISC Processor System

Start-up program is the prerequisite for the operation of RISC processor system, which is part of the whole program block that runs first. Its major function is to make the program start up by choosing a certain memory (SRAM or Flash), to initialize the exception vector table, the EBI data table and the interruption table, to implement address remapping, to set up the stack of processor modes, to allocate the code and data space in the main program, etc. Its operation modes can be put into three categories:

A. System Starts from SRPM and Runs on It

After the system is powered on, the program is downloaded to SRAM through a JTAG simulator, the system starts to implement the start-up program and initialize it from SRAM. Based on the nature of SRAM, it can be known that when the system is powered off (or reset), the codes and data stored in SRAM will be lost, which must be re-downloaded for the next operation. This mode is easy to operate, which can conduct real-time debugging and make it possible to see the variations of parameters such as variable, array, stack, register, etc. Besides, its downloading speed is very fast. In experiments, the debugging of various programs is conducted in this start-up mode.

B. System Starts from Flash and Runs on It

The start-up mode of the system from Flash is very similar to that from SRAM. But there is one difference, that is, after downloading the program to Flash, the program starts from Flash and then operates on it. Based on the nature of Flash, when the system is powered off, the data stored in Flash will not be lost. After disconnecting the simulator, the program can operate automatically provided the system is powered on or reset. Although this mode can realize self-starting, the operating speed of the program on Flash is rather slow, making it hard to bring the computational performance of the RISC processor into full play.

C. Program Starts from Flash and Runs on SRAM

The read speed of Flash is generally dozens of ns, while that of SRAM is about 10 ns. Thus, to improve the operating speed, the program must run on SRAM. Data will be lost when SRAM system is powered off, while it is still kept when the Flash system is powered off. In light of this, the program needs to start from Flash. Thus, it can be certain that the operation mode of the system can be specified as follows:

Program and data are stored in Flash, and start from Flash when the system is powered on. Then the program and read-only data are copied into SRAM. The processor reads the data and instructions in SRAM, which run on SRAM as well.

As for hardware, corresponding configurations are also needed. The start-up mode of the RISC processor depends on whether the level of BMS pin is high or low and whether there is internal ROM or external SRAM. Since AT91R40008 extends its external memory Flash and uses NCS0 signal gating, the BMS pin is thus set to low level, which can help the system select and implement the reading instructions of Flash through NCS0 after the system resets.

(2) Main Program of Star Identification and Attitude Establishment

This part of the program is the core of star sensor and also the major function of the RISC data processing circuit. It mainly consists of three parts: full-sky star identification module, tracking module, and attitude solve module. These modules are first developed and simulated on PC, then implemented to the RISC processor platform, and finally fixed in Flash memory.

Star sensor operates in two modes: Initial Attitude Establishment Mode and Tracking Mode. When star sensor is just launched into orbit, it is Lost In Space. It will resort to the full-sky star identification program to identify star images and search the full sky for the matching guide star. Once the match is successful, star sensor will enter into Tracking Mode. At this time, the position of the celestial area pointed at by boresight has been obtained. Thus, it will be able to identify the measured stars very quickly and, at the same time, establish and output the attitude. During the tracking, if the number of stars in the FOV is too small, the accurate attitude cannot be determined and star sensor will then resort to the full-sky star identification program again. In practical application, most of the time star sensor operates in Tracking Mode and outputs the results of attitude establishment after finishing processing each frame.

(3) Time Testing Program

To realize the 10 Hz attitude output frequency of star sensor, RISC processor must complete data receiving, tracking and attitude establishment within 100 ms. The real time of the system must meet certain requirements. Thus, the time consumption parameters for program running are also very important. To test the accurate time of the software operation, the timing module inside the RISC processor is used. Through control registers TC1_CMR, TC1_CCR, TC1_SR and TC1_CV, accurate timing can be realized. The dominant frequency of the clocks used by the three 16 bit timers/counters of AT91 chip can be put into five frequency divisions (or external clocks). And the maximum one is MCK/1024. When the main frequency is 50 MHz, the maximum time value that can be measured by the 16 bit counter is $216/(50 \times 1024) = 1.28$ s, and its precision is $1/(50 \times 1024) = 1.953 \times 10^{-5}$ s = 19.53 μs. Thus, the timing and counting functions of the RISC processor can be used to test the time of software operation with high precision.

7.2 Hardware-in-the-Loop Simulation Test of Star Identification

Hardware-in-the-loop simulation test mainly uses the star field simulator to test the process of star identification and star tracking by star sensor [2, 4, 5], verifies the validity of the identification algorithm, and tests its performance at the same time.

7.2.1 Test System Configuration and Test Methods

Hardware-in-the-loop simulation test of star identification focuses on the following steps:

① Function test of star identification and star tracking. Based on the results of star identification and attitude output, whether the full-sky star identification and star tracking are correct is judged.
② Time test of the full-sky star identification. The time used for identifying one star image when there is no prior information for the test.
③ Test of attitude update frequency. The processing time of each frame of star image is tested in star tracking and the attitude update frequency is calculated.

The test equipment used in the hardware-in-the-loop simulation test includes star field simulator, data processing computer and optical platform. Star field simulator consists of a small LCD screen, an optical collimation system and a star field simulation computer. The LCD screen is connected to the video output of the computer which generates the simulated star field. Since the LCD screen is installed on the focal plane of the system, the light rays emitted from the screen turn into parallel starlight after passing through the optical collimation system, which can simulate the star image from an infinitely distant point. The major technical indexes of the star field simulator are shown in Table 7.1.

Figure 7.7 shows the physical graph of the hardware-in-the-loop simulation test system. The star field simulator and star sensor are installed on the optical platform

Table 7.1 Technical indexes of star field simulator

Precision of angular distance between stars	≤20″
Simulated magnitude	2–7
Parallelism of starlight	≤15″

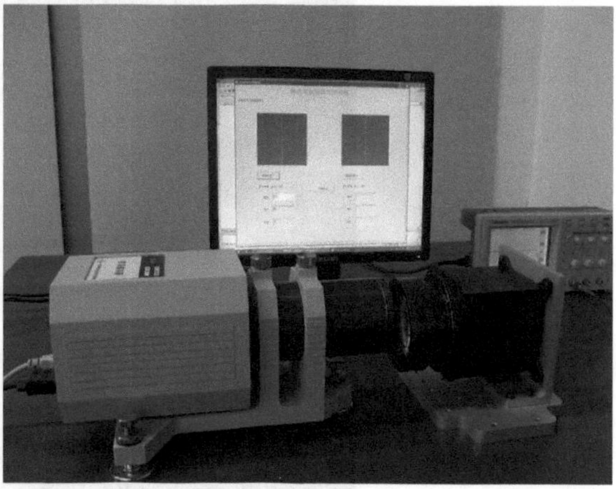

Fig. 7.7 Hardware-in-the-loop simulation test system

and star sensor is connected to the data processing computer by test cables. Based on the motion path of a spacecraft, the star field simulator can calculate the boresight pointing of star sensor and generate a real-time star image. Star sensor is installed and aimed at the lens of the star field simulator. Both of their optical axes should be as parallel as possible, and the joint between them should be shaded to reduce the influence of stray light. Star sensor, through star images generated by star field simulator, simulates the observation of the real night sky and can test the functions of star sensor such as star identification, tracking, and attitude establishment.

7.2.2 Function Test of Star Identification and Star Tracking

As shown in Fig. 7.7, star identification and star tracking tests are conducted by aiming the boresight of star sensor at the star field simulator to verify the accuracy of star identification and star tracking. The FOV of star sensor used in the test is $20° \times 20°$ in size and the star identification algorithm is the modified triangle algorithm based on angular distance matching. The star image simulation program selects stars of 5.0 Mv, and the attitude angle of star sensor is set to 12° (right ascension), 58° (declination), and 90° (roll). Figure 7.8 shows the photographed star image. Since the FOV of the star field simulator is smaller than that of star sensor, only the rectangular part in the middle of the star image is valid. Figure 7.9

Fig. 7.8 Star images photographed by star sensor

shows the results obtained by identifying the photographed star image with star identification software. As shown in Fig. 7.9, the software identifies the photographed image correctly and obtains the correct results of attitude establishment.

Figure 7.10 shows the attitude results obtained by the RISC processor of star sensor through full-sky star identification. Compared with the results of star image processing software and the attitude value set by star image simulation, it can be seen that star sensor correctly identifies the star image simulated by star field simulator.

To further verify the accuracy of star identification in the tracking status, continuous tracking and identification are done, and the star field simulator is set as follows:

The initial attitude angle is 12° (right ascension), 58° (declination) and 30° (roll), with right ascension increasing by 0.2°/s angular velocity, declination by 0.2°/s and the roll angle remaining unchanged. Based on this attitude, dynamic star images are generated to simulate the on-orbit motion of star sensor. Star sensor photographs the

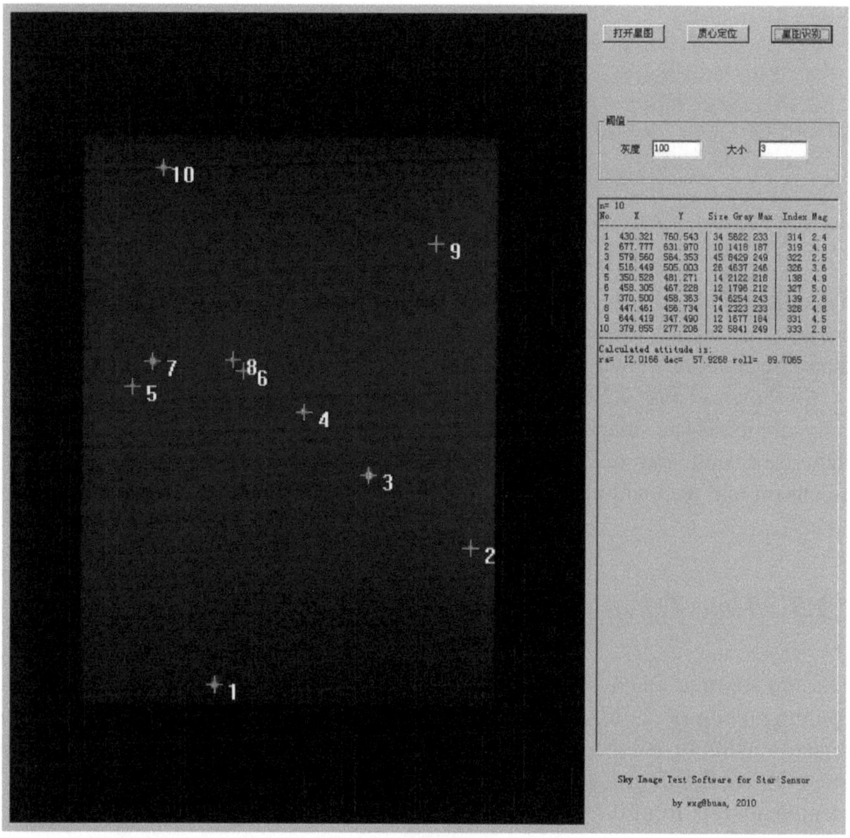

Fig. 7.9 Results of identification through star image processing software

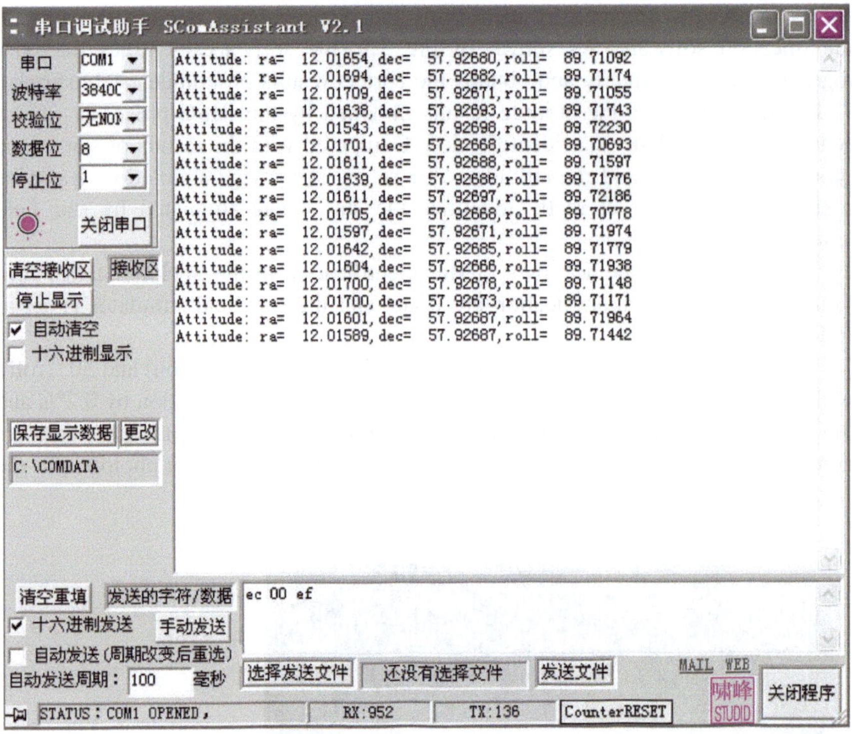

Fig. 7.10 Attitude output results obtained by star sensor through full-sky star identification

dynamic simulated star images, conducts star tracking and outputs the attitude results. The interval of each data output of star sensor is 100 ms and the attitude output results are shown in Fig. 7.11.

As shown in Fig. 7.11, star sensor can conduct stable star tracking of dynamically simulated star images. It can be seen from the attitude establishment results of right ascension and declination angles in Fig. 7.11 that the gradient of right ascension and declination angles correctly reflects the angular velocity (0.2°/s).

7.2.3 Time Taken in Full-Sky Star Identification

The test of time taken in star identification is realized by using the timing and counting modules of an ARM processor. The star field simulator generates 100 star images randomly and the time taken in identifying each star image by star sensor in the FOV is listed in Table 7.2. The time taken by star sensor in full-sky star identification is from 0.18740 s to 1.21544 s, and the average is 0.47832 s.

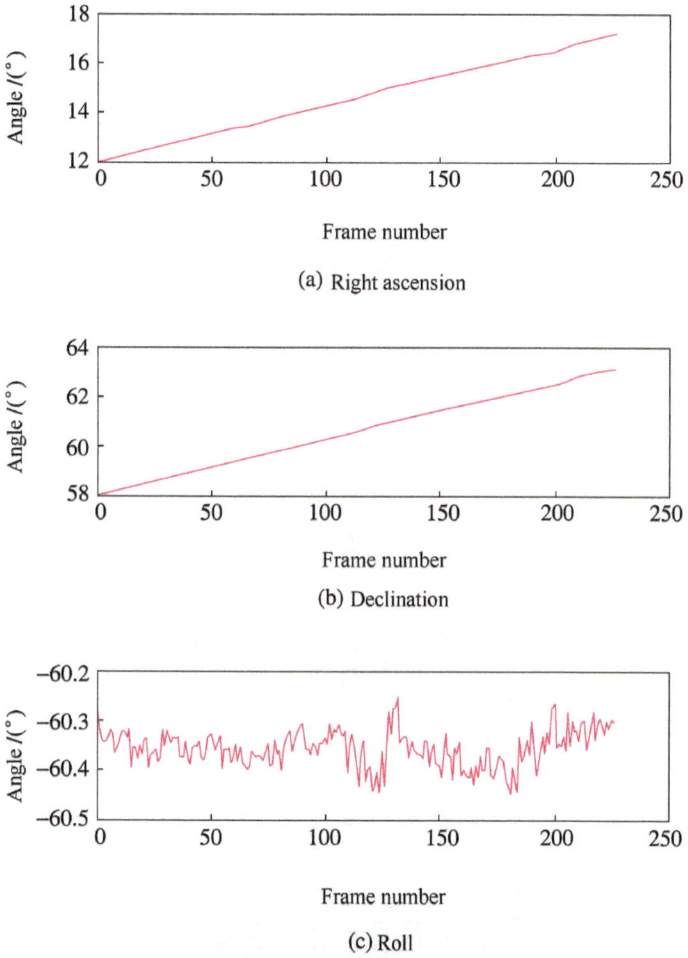

Fig. 7.11 Attitude output results obtained by star sensor through star tracking of dynamic star images

7.2.4 Update Rate of Attitude Data

The way in testing the attitude data update rate is basically the same as that of testing the time taken in full-sky star identification, both of which uses the timing and counting modules of the ARM processor to test the time taken in star tracking.

When star sensor is run in Tracking Mode, ten points are tested respectively when the number of tracked stars is the same, and the time taken in each tracking is recorded (Table 7.3).

Table 7.2 Time taken by star sensor in 100 times of identification

0.33300	0.90202	0.42008	0.43590	0.69418	0.45708	0.29926	0.99366	0.47290	0.23086
0.36716	0.63478	0.41742	0.37654	0.89674	0.37558	0.25880	0.69964	0.26708	0.22958
0.51400	0.40708	1.21544	0.45530	0.66270	0.72104	0.32494	0.30518	0.41712	0.20042
0.35452	0.39614	0.74252	0.79236	0.43496	0.52870	0.58108	0.64760	0.41892	0.37936
0.35452	0.49348	0.44644	0.39704	0.80316	0.27570	0.57750	0.51188	0.41528	0.26636
0.64762	0.30160	0.46904	0.41776	0.98596	0.53854	0.28992	0.44430	0.23040	0.23040
0.42274	0.18740	0.45712	0.68658	0.29410	0.76002	0.52678	0.47202	0.23064	0.22972
0.37452	0.43392	1.06404	0.42514	0.41484	0.49888	0.69688	0.46170	0.22800	0.37784
0.22574	0.62964	0.47268	0.32074	0.41728	0.88876	0.57394	0.68322	0.23200	0.49076
0.55252	0.35612	0.71108	0.64964	0.41606	0.27004	0.53022	0.22960	0.22920	0.39726

Table 7.3 Time taken by star sensor in tracking stars with different numbers

Number of tracked stars	4	5	6	7	8
Tracking time (ms)	0.05516	0.05728	0.06945	0.05972	0.06048
	0.05418	0.05718	0.05546	0.05928	0.07918
	0.05990	0.05704	0.05470	0.05904	0.06912
	0.06958	0.05680	0.05540	0.05894	0.06736
	0.05876	0.05622	0.05528	0.05890	0.06736
	0.05474	0.05512	0.05680	0.06154	0.06746
	0.05450	0.05812	0.05630	0.05796	0.07525
	0.05778	0.05714	0.05608	0.05984	0.07466
	0.05778	0.05710	0.05646	0.05762	0.07464
	0.05620	0.05660	0.05790	0.05770	0.07502
Average value	0.05786	0.056860	0.05738	0.05905	0.07105
Number of tracked stars	9	10	11	12	13
Tracking time (ms)	0.06214	0.06348	0.07102	0.06994	0.08958
	0.06282	0.06318	0.07130	0.07002	0.08670
	0.06126	0.06760	0.07130	0.06982	0.08958
	0.06946	0.08656	0.07128	0.06994	0.08670
	0.06922	0.08366	0.07126	0.06952	0.08674
	0.06926	0.07610	0.07114	0.06972	0.09266
	0.08286	0.07678	0.06912	0.06950	0.09076
	0.08130	0.07706	0.07860	0.08668	0.08954
	0.08134	0.07720	0.07884	0.09656	0.09042
	0.08136	0.07698	0.07850	0.09816	0.09058
Average value	0.07210	0.07486	0.07324	0.07699	0.08933

In Tracking Mode, the maximum number of stars that can be tracked by star sensor is 13, and each tracking can be completed in 100 ms, i.e., the attitude data update rate of star sensor can reach 10 Hz.

7.3 Field Experiment of Star Identification

To test the performance of star sensor more correctly and understand the operational status of star sensor in practical application, the real night sky needs to be observed for further testing. Similar to the hardware-in-the-loop simulation test, the field experiment focuses on investigating whether the functions of star sensor (star identification and star tracking) are effective in the real night sky. The experiment also conducts preliminary evaluation of the star sensor's performance in attitude establishment.

7.3.1 Manual Star Identification by Using Skymap

When the star sensor observes stars in the field, a comparison is usually done between star images photographed by star sensor and those simulated by Skymap software before verifying the full-sky star identification algorithm. Through manual identification, the correct match of measured stars in the star image is determined, providing references for the accuracy of star identification. Figure 7.12 shows the main interface of Skymap.

Skymap is a powerful night sky simulation software, which can provide a wide range of astronomical reference information for astronomers and fans. The main functions of Skymap are as follows:

① Display the night sky that can be observed at any place on the earth between 4000 BC and 8000 AD. The observation scope can be as large as the whole sky, or as small as a tiny region.
② Zoom in or zoom out of the celestial areas that are of interest and rotate the night sky through the use of keyboard or mouse.
③ Display more than 15 million stars and over 200 thousand extended celestial bodies: star cluster, nebula, galaxy, etc.
④ Display the positions of the sun, the moon, and the major planets with a margin of error less than 1″.

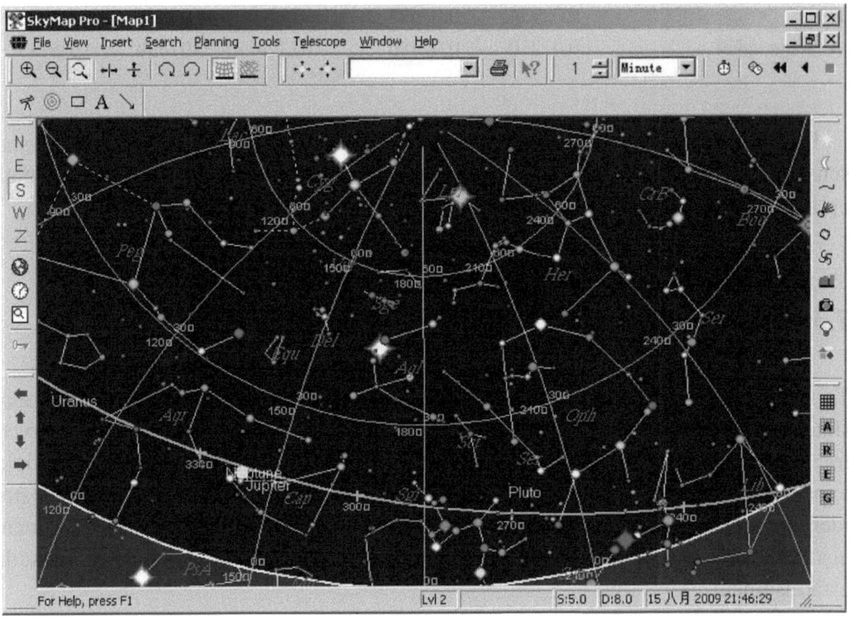

Fig. 7.12 Main interface of Skymap

⑤ Display the names of 88 constellations, their shape connection and all the known asteroids and comets (including a database of over 11,000 asteroids and comets).

⑥ Display the grids and graduation lines of various different coordinate systems such as the horizon coordinate system, the equator coordinate system, the ecliptic coordinate system, the galactic equator coordinate system, etc.

⑦ Annotations can be added to the star image. A circular FOV with camera and rectangular CCD can be randomly configured for observation.

Figure 7.13 shows the real picture of the star observation field experiment. And Fig. 7.14 shows the star image photographed when the boresight of star sensor is pointed at Cassiopeia and its surroundings. The FOV of star sensor is 20° × 20° and its exposure time is 200 ms. Figure 7.15 shows the centroiding results of photographed star images. The specific procedures of centroiding are available in Sect. 2.4.

The longitude and latitude of the place for field star observation, the measurement time, the size of the measured FOV, and other parameters are set in Skymap software. The celestial area near Cassiopeia is selected for comparison. Figure 7.16 shows the star image of the celestial area near Cassiopeia generated by Skymap.

Through one-by-one comparison, star spots (marked by cross lines) extracted from the real star image can find their corresponding stars. The comparison results are shown in Table 7.4. It can be seen that each of the measured stars in the measured star image find its corresponding guide star through manual identification.

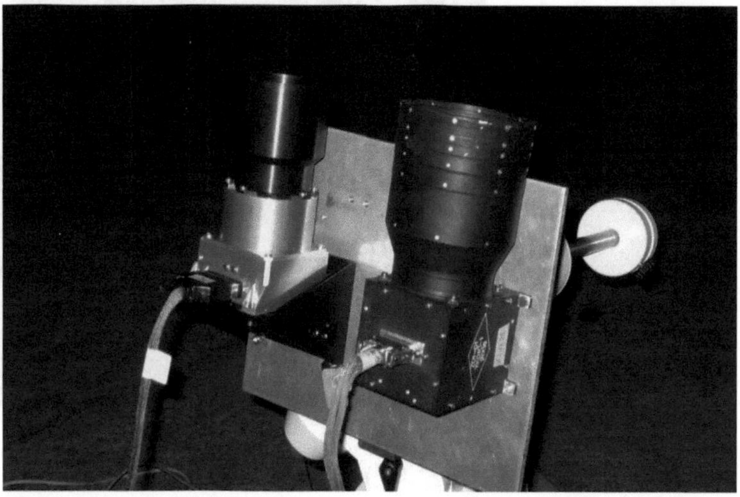

Fig. 7.13 Star observation physical graph of the star sensor in the field

Fig. 7.14 Star image photographed when the boresight of star sensor is pointed at Cassiopeia and its surroundings

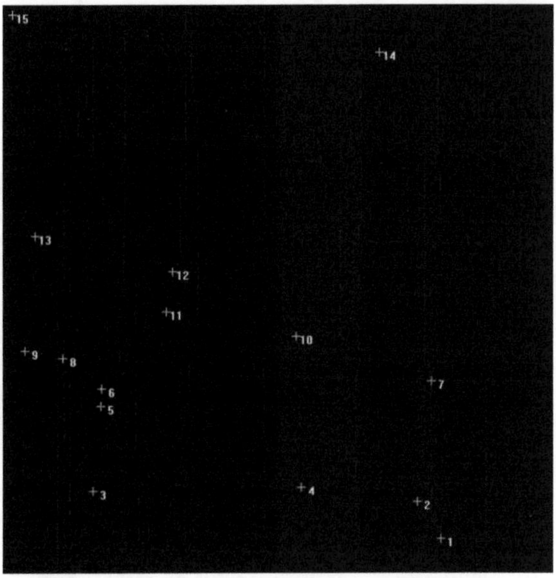

Fig. 7.15 Centroiding results of the photographed star images

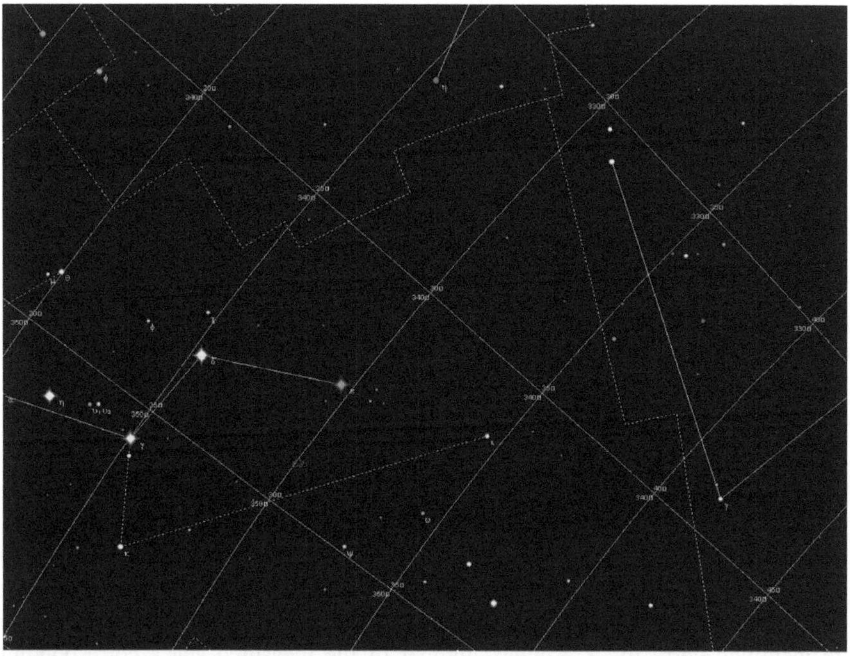

Fig. 7.16 Star image of the corresponding celestial area generated by Skymap

Table 7.4 Comparison results

Index number	Name	Visual magnitude	Coordinates
1	Cassiopeia 50	3.95	Right ascension 02 h 04 m 28.44 s Declination +72° 28′ 45.3″
2	Cassiopeia 48	4.49	Right ascension 02 h 02 m 56.96 s Declination +70° 57′ 55.0″
3	Cassiopeia kappa	4.17	Right ascension 00 h 33 m 39.94 s Declination +62° 59′ 56.1″
4	Cassiopeia psi	4.72	Right ascension 01 h 26 m 46.67 s Declination +68° 11′ 35.2″
5	BU 497	4.80	Right ascension 00 h 53 m 46.45 s Declination +61° 11′ 22.0″
6	Cassiopeia gamma	2.15	Right ascension 00 h 57 m 25.07 s Declination +60° 46′ 56.3″
7	Cassiopeia iota	4.46	Right ascension 02 h 30 m 03.27 s Declination +67° 27′ 21.5″
8	Cassiopeia upsilon2	4.62	Right ascension 00 h 57 m 21.82 s Declination +59° 14′ 47.2″
9	Cassiopeia eta	3.46	Right ascension 00 h 49 m 45.97 s Declination +57° 52′ 57.1″

(continued)

Table 7.4 (continued)

Index number	Name	Visual magnitude	Coordinates
10	Cassiopeia epsilon	3.35	Right ascension 01 h 55 m 14.98 s Declination +63° 43′ 45.6″
11	Cassiopeia delta	2.66	Right ascension 01 h 26 m 34.47 s Declination +60° 17′ 53.5″
12	Cassiopeia chi	4.68	Right ascension 01 h 34 m 42.15 s Declination +59° 17′ 37.6″
13	Cassiopeia	4.34	Right ascension 01 h 11 m 48.42 s Declination +55° 12′ 50.7″
14	Cassiopeia eta	3.77	Right ascension 02 h 51 m 33.61 s Declination +55° 56′ 40.5″
15	Cassiopeia 51	3.59	Right ascension 01 h 38 m 42.63 s Declination +48° 41′ 19.5″

7.3.2 *Function Test of Star Identification and Star Tracking*

The real star images photographed by using the modified triangle algorithm based on angular distance matching (introduced in Sect. 3.2) are identified. The results are shown in Fig. 7.17. It can be seen that the measured stars in the measured star image are identified correctly. The index number of the measured star's corresponding matching guide star can be found in the GSC. Through this index number, the SAO index number of the star can be searched out from the SAO J2000 star catalog. Both of the above results are in line with those of the manual identification of Skymap. It is noticeable that there are some subtle differences between the magnitude information in the SAO star catalog and that in the Skymap star database.

Through the identification results, the star probe sensitivity of star sensor can be estimated. It can be seen from the identification results in Fig. 7.17 that stars of 5 Mv can be measured by star sensor. In fact, if the gray threshold (set to 50 in Fig. 7.16) is lowered, a larger number of fainter stars can be extracted.

In an actual star observation test, a large number of photographed star images go through full-sky star identification test. It has been tested that 100% identification rate can be obtained when the number of measured stars in the FOV is bigger than four.

In star observation field experiments, specific tests have been done when there emerge interfering stars. Figure 7.18 shows the identification result of a photographed star image at Nanshan Observatory, Xinjiang (longitude and latitude: N43°28′18″, E87°10′45″) at 8 a.m. on December 31, 2008. The FOV of star sensor is 20° × 20°, and its exposure time is 100 ms. As can be seen from the results, the

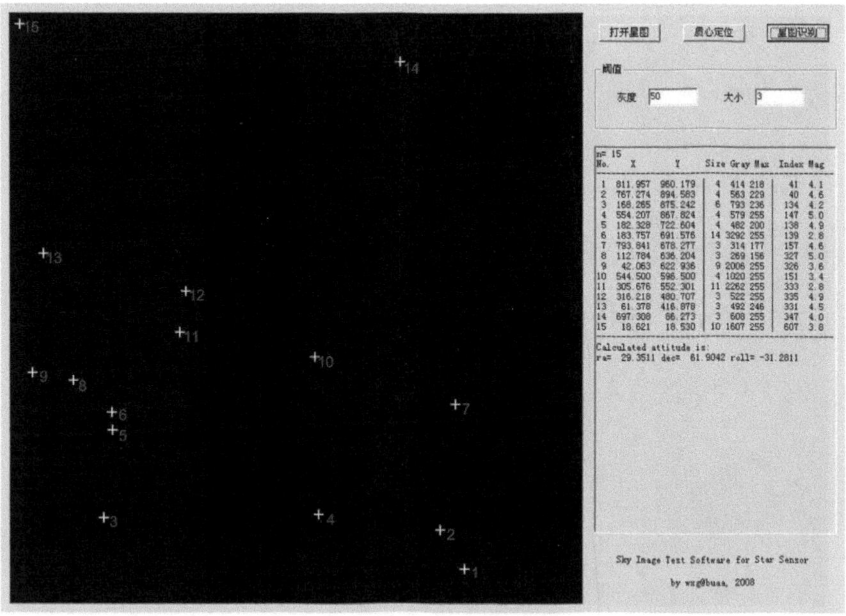

Fig. 7.17 Results of star identification by using modified triangle algorithm based on angular distance matching

bright star marked 4 is not identified correctly (Index and Mag are marked by −1 and 99.0, respectively.) The star image obtained at this place and time is simulated by Skymap software, as shown in Fig. 7.18. It can be seen from Fig. 7.18 that the star marked 4 is Saturn. The existence of Satum in the FOV does not affect the correct identification of other stars or the correct output of the attitude establishment result. Figure 7.19 shows the star image of the corresponding sky zone generated by the Skymap.

In Tracking Mode, the accuracy of star tracking is verified by continuously outputting the attitude. Figure 7.20 shows the result curve of attitude output after continuous star tracking of 3000 frames. As can be seen from the curve, star sensor can conduct stable tracking and attitude output in the process of field star observation. The attitude variation gradient of its right ascension angle reflects the earth's rotation velocity.

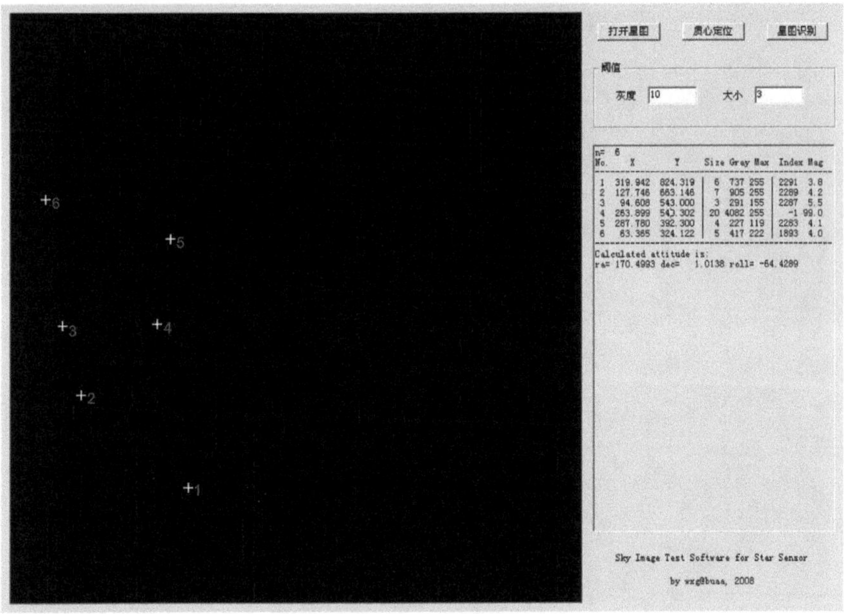

Fig. 7.18 Identification results when there are interfering stars in the measured star image

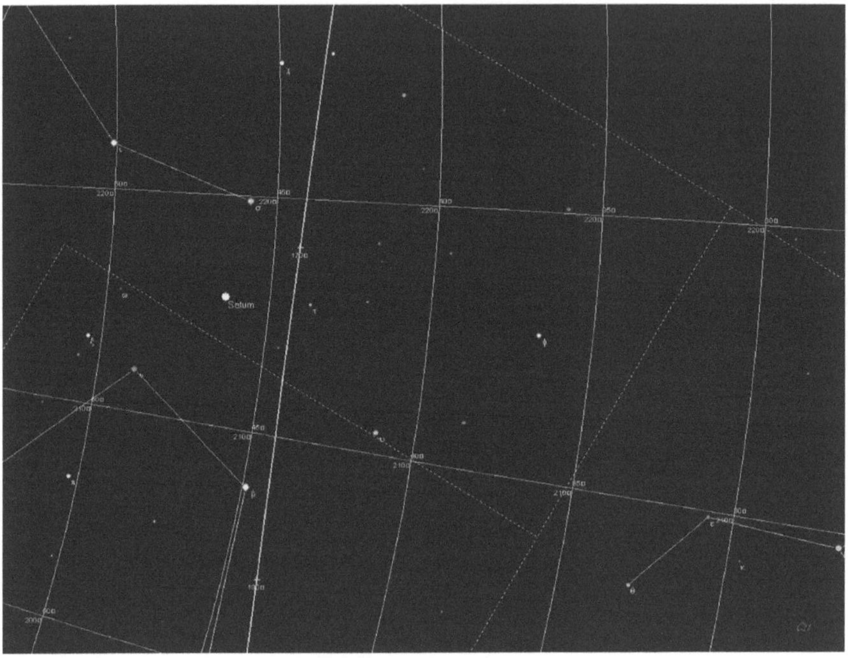

Fig. 7.19 Star image of the corresponding celestial area generated by Skymap

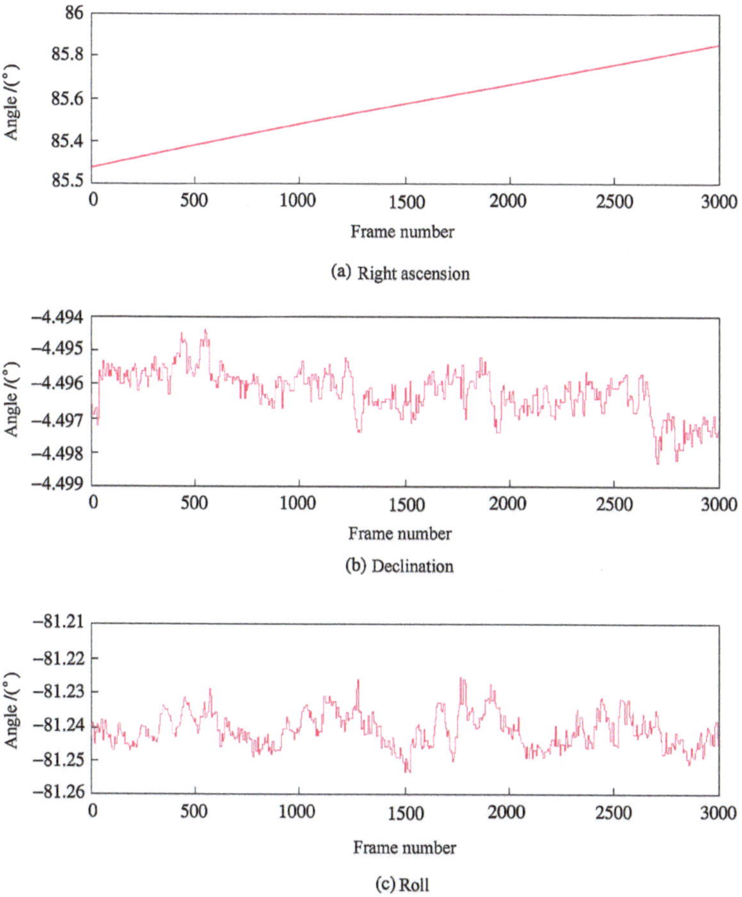

Fig. 7.20 Result curve of attitude output of continuous star tracking of 3000 frames

References

1. Yang J, Jiang J, Zhang G, Li S (2005) RISC technique in star sensor, Opto-Electron Eng 32 (8):19–22
2. Yang J (2007) Star Identification Algorithm and RISC Technique, Doctoral Thesis of Beijing University Aeronautics and Astronautics, Beijing, pp 62–96
3. AT91R40008 electrical characteristics, http://www.atmel.com
4. Yang J, Zhang GJ, Jiang J, Fan QY (2007) Semi-physical simulation test for micro CMOS star sensor. International Symposium on Photoelectronic Detection and Imaging Beijing, China, SPIE, vol. 6622
5. Wei X, Zhang G, Fan Q, Jiang J (2008) Ground function test method of star sensor using simulated sky image, Infrared Laser Eng 37(6):1087–1091

MIX
Papier aus verantwortungsvollen Quellen
Paper from responsible sources
FSC® C105338

If you have any concerns about our products,
you can contact us on
ProductSafety@springernature.com

In case Publisher is established outside the EU,
the EU authorized representative is:
Springer Nature Customer Service Center GmbH
Europaplatz 3, 69115 Heidelberg, Germany

Printed by Libri Plureos GmbH
in Hamburg, Germany